MacBride, James Douglas

A Handbook of Practical Shipbuilding

Inktank publishing

MacBride, James Douglas

A Handbook of Practical Shipbuilding

Inktank publishing, 2018

www.inktank-publishing.com

ISBN/EAN: 9783747750032

A HANDBOOK
OF
PRACTICAL SHIPBUILDING
WITH A GLOSSARY OF TERMS

BY

J. D. MacBRIDE

MEMBER, SOCIETY OF NAVAL ARCHITECTS AND MARINE ENGINEERS
FORMERLY SUPERINTENDENT IN HULL CONSTRUCTION
HOG ISLAND SHIPYARDS

SECOND EDITION, REVISED AND ENLARGED

204 ILLUSTRATIONS, 2 FOLDING PLATES

NEW YORK
D. VAN NOSTRAND COMPANY
EIGHT WARREN STREET
1921

PREFACE TO SECOND EDITION

"PRACTICAL SHIPBUILDING" deals primarily with **Shipbuilding,** i.e., the work on and around the shipways. The allied trades in the shops (machine, pattern, joiner, blacksmith, etc.) are distinctly separate trades, as they are also employed in many other kinds of commercial enterprises. Therefore the details of that work do not interest the reader of this volume, rather, only the work which specially pertains to ship construction.

The first edition has been thoroughly overhauled. No repairs were made, but considerable was done in the way of additional information under some of the former chapters and new material has been added in the form of Chapters XII and XV.

Shipbuilding is an extensive industry and the more one knows of the immense amount of detail in it, the more one realizes that such is really the case. It is impossible to go into details in a volume of this size, therefore it is the endeavor of the writer to deal only with those everyday things which the men on the shipways should know.

In building a small book of this kind it is more difficult to know what to leave out than it is to know what to put in it. It is like a "camouflaged" ship, the viewpoint is different, according to each person who looks at it.

It is hoped that the voyage of the **Second Edition** will be as successful as that of the **First,** which received cordial greetings in many harbors; with that in mind the anchor is weighed and "departure" is recorded.

J. D. MacBride.

December, 1920.

PREFACE TO FIRST EDITION

This handbook on the practical construction of a modern standard cargo steamer has been written in answer to some of the many questions which from time to time have been asked by men who have been working under the supervision of the writer, and is intended to fill the need for a guide to the new men starting in shipyard work.

During many years' experience in shipyards on all types of marine construction from the smallest of torpedo boats of years ago to the mighty war vessels and merchant ships of to-day the writer has been associated with the men who are to-day supervising the wonderful shipbuilding program on which this country is engaged, and he has endeavored to embody the results of this more than twenty years' experience in the book.

In the present struggle on land and sea, when everyone must use all his might, this manual of the necessary steps in fabricating and assembling of vessels has been prepared with the hope that it will prove a help to some of the many thousand men who must come into this industry from other trades.

Team work is necessary in all lines, sports and commercial enterprises alike, but in none is it more essential than in an enterprise where so many different trades are all working together, and the results of their labor must be co-ordinated to effect a satisfactory conclusion.

The riveters must follow the reamers and the reamers must follow the bolters-up, etc., therefore, it is necessary that all the men working on the ship should have an intelligent understanding of the general plan.

J. D. MacBride.

November, 1918.

TABLE OF CONTENTS

PRACTICAL SHIPBUILDING

CHAPTER I

Shipyard Organization

The shipbuilding business is like any other in that its executive and financial departments are separate from the manufacturing or fabricating branches of the organization. As this volume is not concerned with the Economics of Shipbuilding, but with ship construction, we need not burden ourselves with descriptions of the duties of officials like the President, Treasurer, etc., whose functions do not bring them directly in touch with the actual work of putting a ship together.

In a modern shipyard the General Manager has charge of both Hull and Engine Departments, but he is the only connecting link, the two branches working quite independently of one another.

The Engine Department has a drafting room in which the designs and all detail drawings for use in the yard are made. Much of the work is made up of castings which are shipped in from outside firms according to plans furnished them and the shipyard is interested in making the rough casting over into a finished product ready for installation in the ship. Nearly all large shipyards have a machine shop with the necessary lathes, planers, shapers, etc., to do the work, with ample overhead crane facilities to handle the heavy weights.

The Machine Shop often has the heavy work on the floor, with galleries above to carry the smaller and lighter weight machines and fittings where much of the " bench work " is done on the small brass castings. When the heavy lathes, planers, etc., are at one end, on the floor, the other end of the building is devoted to erection work. The engine bed is laid down and thoroughly bolted, and the engine then put together piece after piece. This is done to be sure it will fit when carried to the ship, as this assembly work can be done while the ship is still under construction on the shipway. Any necessary changes can be made at this time without delaying delivery of the ship to the owners. After the whole engine has been assembled, it is taken down and sent to the ship as soon as the Engine Room is ready to receive it. This work is done for both reciprocating and turbine engines.

This also applies to such parts as propellers, propeller shafts, thrust blocks, etc. The condensers, auxiliary pumps, and fittings are often bought from an outside firm all ready for installation in the ship.

The Boiler Shop is the other large portion of the Engineering Department and in it the boiler work is all laid out. The equipment of this shop varies with the type of boilers that the shipyard is using, whether " Scotch " or " water-tube " types.

When the " Scotch " boilers are used the heavy boiler

plates are curved to the required circular shape by heavy, upright rolls. These plates are assembled, the furnaces, stay-rods, etc., fitted and after being riveted, caulked and tested with a water-pressure, they are set outside the shop until the Boiler Room is ready for them. "Water-tube" boilers are made up of many small parts and are often put together on the ship for the first time. The parts are light in weight and easily handled.

The Hull Department, has a drafting room separate from that of the Engine Department, in which the hull structure is laid down and detail drawings made of the different parts of the hull, such as shell, deck, and bulkhead plating; also all details for the hull fittings.

Each detail plan of the steel plating of decks, bulkheads, etc., gives all necessary information regarding the thickness of plates, the symbol whereby it may be traced from the Steel Yard, and dimensions giving all sizes. Where special riveting is required this is also shown in detail, but most of it can be described by notes, giving the diameter, spacing and type of head and point.

These plans show where all the angles are placed, all butts in plates and angles and all information necessary for assembling the whole of that part of the ship.

The Steel Yard is the place where the material for the hull is stored until it is ready for use. The various plates being placed together according to their destination on the ship, as shown by the marks painted on them by the steel makers.

All plates are stowed on edge so any of them may be taken out without disturbing all the others. They are held together and kept from falling by means of a rack which is sufficiently wide for the Yard men to lay the plates over at a small angle so they can look over the plates in order to find the symbols which indicate where each is to go when it is ready to send to the ship.

The angles and channels are stowed separately and so placed that they can be easily identified when needed.

All material, being heavy, is handled by cranes either over head or of the traveling type.

Bookkeeping for all this material in a large yard is a huge task and must be done carefully or the work of the yard would at once come to a standstill. Each piece of plate or angle must be stowed so it can be located and taken out from a large pile of other steel on a few minutes' notice.

The Mold Loft is a long, narrow room with an exceptionally smooth floor on which the outlines of the various parts of the hull are laid down and drawn full size. The floor is divided up for various parts of the work of laying off the ship. One part is devoted entirely to the "Body Plan" and another exclusively to the heights and half breadths of the decks and water lines. Usually in one corner there is an office for the man in charge with facilities for keeping his blue prints, "Table of Offsets" and other data furnished by the Hull Drafting Room.

A Complete Set of "lines" are furnished the Loftsmen and from these may be obtained all the information necessary for laying out any portion of the vessel. They comprise: (1) The "sheer plan" or heights, which is a longitudinal section through the ship showing the heights of all decks above the keel, bow and buttock lines. (2) The "half-breadth" plan shows the outline of all deck and water-lines. (3) The "body plan," is a series of cross-sections, or curves of the frame lines of the ship for its entire length.

At one end of the Mold Loft floor, or often a temporary one laid over it, is located the "Scrieve Board." This is usually painted black before any work is done on it, because lines scrieved (cut) into it stand out in better relief and are easier to see when "scaling" (measuring) dimensions.

Mold material is thin wood, about ⅛ in. thick and about 4 to 6 in. wide which is tacked together to suit the full-sized shape of the plates and on this is marked the exact shape of the plate and the location of all rivet holes. Heavy cardboard, being much lighter and easier to handle than wood, is often used for molds, but of course will not stand quite as much handling. Each mold is marked with the same symbols as the plate and is then sent down to the Shipyard.

When a mold arrives and work is ready to begin on a plate, it is taken from the rack and sent to the ship fitter, somewhere near the ship shed.

Here the ship fitter lays the mold on the plate, marks off the outline, also each rivet hole. When ordering the plate in the Drafting Room an allowance of 1 in. on the length and ½ in. on the width is added extra on the " dead flat" (perfectly straight and flat) plates, while a little larger allowance is made for angle bars, so the fitter has to be careful and be sure that he is quite correct.

From here the plate is carried over to the Ship Shed.

The **Ship Shed,** or **Plate and Angle Shop** as it is often called, is a long, low building in which is done the fabricating of the material.

There are usually at least two furnaces, sometimes three, at one end of the shop. The plates are handled down one side of the building; the angles down the other side.

Plates are spoken of as "dead flat," "half furnaced," or "full furnaced." Plates which are dead flat, without any curvature whatever, are at once laid off by the ship fitter. The outline is laid off, location of all rivet holes and type of hole (countersunk or not) and identifying symbol is painted on the plate and it is then ready for the "shears" and "punch."

Plates that are to be half or full furnaced are shoved into the furnace and heated until red, when they are hauled out onto the "bending slab" and flanged or bent according to iron molds which have been prepared previously.

Such plates as do not require sharp bends are put through the "bending rolls." These have one roller on top and two below at adjustable distances from each other. After one edge of a plate has been caught between them the rolls work it back and forth, rolling it one way and then the other until any desired curvature is given to the plate. Some plates have "sny" (curvature in both directions, across and lengthwise) and it is rolled into them by entering the plate diagonally.

Some yards have another set of rolls, with two on top and three on the bottom row which are used for straightening plates received badly bent and out of shape.

After these plates have been furnaced or rolled to suit the required shape, as checked by the mold, they are then taken in hand by the ship fitter, who adjusts his template and lays out the size and rivet holes, as for the flat plates.

In some of the newer ship sheds the material is transported by overhead traveling cranes of about five tons capacity but in many of the older yards the sheds are equipped with narrow-gauge rails and small flat cars. When it is necessary to move at right angles, small turntables are located so the cars can be swung around.

After the plate has been rolled flat or to the curved surface required, it is carried to the **Shears**, which are large, heavy cutters, sufficiently long to take the length of the average plate at one time, where it is cut to the lines as marked from the mold by the fitter.

The Punch is the next step in the program. The plate is laid on a series of small wheels which are set in a frame, allowing them to face in any direction. These wheels are close up to the punch and the operator on the punch has one man, or more, to assist in pushing the plate in the desired direction. The punch man stands at his machine and moves the plate until the rivet marks, as set off in small circles with white lead, are directly under the punch, when he operates the machine and the rivet hole is punched out, the "punching" dropping onto the ground to be gathered up later on and sold for scrap.

After the plate has been sheared and punched, if it is to be an "outside" plate it must be countersunk. (Should it be an "inside" plate it is passed right on to the ship.) **The Countersink** is a drill with an entrance of 30 degrees, as a general rule, and it is drilled into the rivet hole until the straight part of the first hole has been cut down to

$\frac{1}{16}$ in. in depth. Thus, every countersunk hole should have a straight length and then flare out at the bevel, or angle of the countersink. This is done to be sure that the rivet will " draw " the two plates, or plate and angle, together.

The plate is now ready for the ship.

The Bending Slab for the frames is usually located under the Ship Shed roof, and is described under "Frames" in Chapter VII.

The Blacksmith Shop is an important part of the shipyard equipment, as all the small work such as rail stanchions, boat davits, mast and boom fittings, etc., are turned out here. The blacksmiths must be good workmen, as much of the work is required to be close to the dimensions given on the blue prints from which they work.

One portion of the shop is turned over to "tool dressers" who make and keep in repair the hand and machine tools which are used by the chippers, caulkers, etc., on the shipways.

In some yards a heavy hammer is located in one end of the shop for use when making "drop forgings." These are generally small fittings of which a large quantity is required. A die is made in two parts, hot metal dropped in it and then the hammer forces the die together, forming the fitting.

Until recently it has been necessary to maintain an "angle smith" force for working the water-tight angles on floors, water-tight angle collars and staples on decks but much of this work is now done by the electric welders.

The Galvanizing Shop is necessary as much of the material from the Blacksmith Shop has to be galvanized to avoid rust from exposure to salt air and water. This is also true of some of the piping in certain parts of the ship and as much of it has to be bent it is bought black, bent to shape, cleaned in a dilute acid wash, brushed, dried and then galvanized.

This shop is equipped with pickling and galvanizing tanks. The number and size of the tanks varies according to the size of the ship yard and the amount of work required. For instance, one yard which is building torpedo boat destroyers for the Navy requires many of the plates to be galvanized. This yard has a large installation, with tanks big enough to take a large plate at one time.

The Joiner Shop gets out all the inside wood work for the ships, such as berths, tables, lockers, etc., in the shop and when the ship is far enough advanced it is then carried out and placed on board. Some pieces which may be too large to pass through openings in the ship are " knocked down " and then put together again in the ship.

The Pilot House is often built in the Joiner Shop and then skidded down to a crane or " shear legs " and lifted aboard and into place on the Bridge.

The Pattern Shop makes all patterns for ship and deck fittings such as hawse pipes (rudder and stern frames are made near the place of manufacture, at the foundry), chocks, bitts, etc. As the deck fittings are of a standard type the patterns for them may be used many times, and such patterns are carefully stored for future use; but hawse pipes have to be made to suit the shape of the bow of the ship and as this is different with each new design they are destroyed after having served their purpose.

Some shipyards have a **Brass Foundry** for casting small deck fittings and others used in the Engine Room, and elsewhere. Steel castings are generally bought from outside firms, which are furnished with the company's patterns for the work.

The Pipe Shop is one of the important shops in the yard, as it has to furnish all the piping for the Engine and Boiler Rooms, Bilge and Ballast Systems, Fresh Water, Sanitary and all other piping throughout the ship. This work includes all kinds of pipe fitting, flanged and threaded, in all sizes, of brass, copper, and steel.

This shop also has a department of sheet iron work, and takes care of all the ventilation piping, cowls on deck, etc. It gets out the material in the shop and has a force on the ship installing it as soon as the ship is in an advanced stage to permit this kind of work.

The Acetylene Department has lightened the labor in recent years in the shipyards as much of the burning out of holes, cutting and trimming of plates, and angle bars, and work of that kind, which took a lot of time, partly because the material had to be brought to the machine which could do it, can now be quickly done by one man. Some yards use it for welding water-tight staples in angle bars, water-tight corners, etc., but not for any part of the structure where strength is required. The Acetylene Department is now a fixture as it has passed the experimental stage on ship work.

The acetylene workman generally uses one tank for oxygen about 9 in. diameter and 4 ft. long, and one tank for acetylene about 12 in. diameter and 3 ft. long

The oxygen and the acetylene tanks are made especially for this work. The operator uses these tanks laying on their sides, hitched together by a Regulator which governs the flow of the gases. Three-eighth inch diameter rubber pipe is used from tanks to the work, in 50-ft. sections, often three 50-ft. pieces being hitched together. For cutting work a straight line and a circle cutter are used. The first can be guided by another plate for a marker and the circle cutter has a beam with a pivot for the center, thus cutting a true circle. Thickness of the material to be cut determines the size of burning " tip." These vary from No. 1 to No. 5 for thin plates up to about 8 in. thick. The welding is done by the " welding torch," and welding rods. These rods are $\frac{1}{4}$ or $\frac{3}{8}$ in. in diameter and are fused by the heat, so they drop onto the material to be welded, which has become hot, and form a weld. Welding tips vary from No. 1 to No. 12 for light to heavy material to be welded.

Without the **Air Compressor House** a modern shipyard would have to shut down, as a greater part of the work formerly done by hand is now done by an " air machine," whether it is a riveting " gun," a reamer, chipping hammer or any of the many different air tools now so commonly used.

In the " old " days (a few years ago) two men did the riveting together, one man striking the rivet as the other was raising his hammer for the next blow. The " holder-on " would use a heavy hammer to back-up the head of the rivet. Now he has an air machine which he can set against the rivet head, place the leg against a wood backing or other support, and let the machine take the blow from the riveter on the other end of the rivet.

Compressed air is carried all through the yard to the various shipways and shops, in steel pipes placed under ground, and is carried at a high pressure because of the many men using it at the same time.

An hydraulic riveting machine often called the "Bull Machine" is installed in a part of the yard which is easily accessible and all the riveting work, such as floors and some foundations, which can be done "snap" (with a "button" point) is done here before sending the material to the ships. This machine is fixed in a permanent position and the material is handled around it by means of a small, electric crane. In this way much of the steel can be made ready and assembled before the ship is in a condition for it to be installed, thus saving much time credited to construction of the ship.

The Tool House is one of the busiest places in the modern shipyard, as it is there that all the tools for work on the ships are stored, repaired and given to the men for each day's work.

This department generally has a building with many windows, so that it is possible for a number of men to be served at one time. Each different trade has a separate window for its use.

The air tools used in the shipyard are expensive and difficult to keep in good repair, as many of the workmen are careless in the use of them. A number of men are busy at all times in the repair and up-keep work in the Tool House, taking the pneumatic hammers, motors, grinders, etc., apart, carefully greasing them and replacing any worn parts. As the workmen on the ways never stop to do this very necessary work, it means that much care must be exercised by the up-keep men in watching the condition of these tools.

In addition to the air tools, all the other tools described in Chapter II are kept in the Tool House for use of the workmen as needed.

In order to keep track of the many tools out of the house at one time a system of brass checks is used. Small tool boxes are issued to the men so they are able to keep some of their belongings on the ship during working hours, without having to leave their work and lose time going for them. As soon as a man is employed he is given a quantity of numbered brass checks and his name and number recorded.

When a tool is needed it is given to the workman at the window of the Tool House, in exchange for one brass check. As soon as he is through using the tool and returns it to the Tool House his brass check is given back to him. In this way it is possible for him to use the checks over and over again, as often as he needs tools.

To facilitate handling in the Tool House all the tools are sorted according to the different kinds (the air tools being stored in wood racks so they are easily reached), a hook just below the space being used to hang the brass check of the workman on when the tool has been given out for use.

This scheme furnishes a simple system of keeping track of tools and avoids bookkeeping and the liability of errors.

As the workman, when leaving the employ of the company must receive a "tool house clearance slip" showing that his account with them is OK, before he can receive his money, he remembers to keep good care of his checks during his term of employment.

Directly after launching, the ship is towed to the **Fitting-out Wharf**. Here some means is provided, either with a "shear-legs," or some form of traveling crane, for lifting in the heavy weights, such as engines and boilers, which were not placed before the launching of the ship. A railroad track is carried along the wharf so that much of the material yet to be installed can be sent down and swung aboard.

Shear-legs are two spars, formed of steel plates, large at the middle and tapering at the ends, which meet at the upper ends and have the bottom ends wide apart. The foot of the spars rest on the edge of the wharf and the tops are tilted out over the water so they will plumb the center line of an ordinary ship, and are held with back-stays of steel wire rope. With a "fall" of steel rope and blocks and a winch on the wharf all material can be easily raised from the wharf, swung out over the ship and lowered on board.

The traveling cranes are usually on a track on the wharf with an arm overhanging the water.

The following table gives a general idea of the manner in which the work of shipbuilding is sub-divided. Each step follows the other in logical sequence, although at certain stages of progress in the work a number of different operations are carried on separately up to the time for assembly.

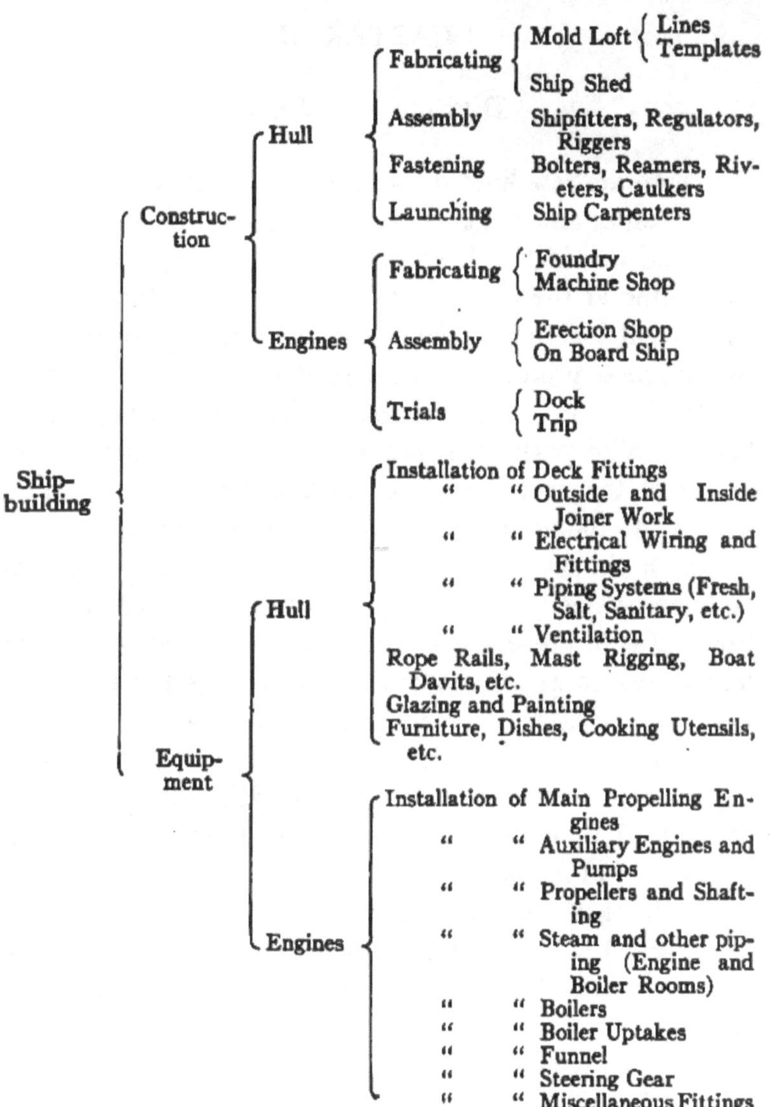
Ship-building
Construction
Hull
Fabricating
Mold Loft
Lines
Templates
Ship Shed
Assembly
Shipfitters, Regulators, Riggers
Fastening
Bolters, Reamers, Riveters, Caulkers
Launching
Ship Carpenters
Engines
Fabricating
Foundry
Machine Shop
Assembly
Erection Shop
On Board Ship
Trials
Dock
Trip
Equipment
Hull
Installation of Deck Fittings
" " Outside and Inside Joiner Work
" " Electrical Wiring and Fittings
" " Piping Systems (Fresh, Salt, Sanitary, etc.)
" " Ventilation
Rope Rails, Mast Rigging, Boat Davits, etc.
Glazing and Painting
Furniture, Dishes, Cooking Utensils, etc.
Engines
Installation of Main Propelling Engines
" " Auxiliary Engines and Pumps
" " Propellers and Shafting
" " Steam and other piping (Engine and Boiler Rooms)
" " Boilers
" " Boiler Uptakes
" " Funnel
" " Steering Gear
" " Miscellaneous Fittings

CHAPTER II

Division of Work

There are few lines of business in which so many different trades and kinds of work are carried on at the same time and in which it is so necessary that they shall all be continuing at the proper rate to be ready for the different parts to fit together at the proper time in order to turn out the complete product as in the modern ship yard.

This result is accomplished by close co-operation between all departments and in an exact knowledge of the extent of the work each is to do and the time required for delivery to the department for which it was made.

In the larger shipyards material is "routed" so it will be known when each special piece will be ready for installation. For instance a wrought-iron rail stanchion must be forged, drilled at the machine shop for pins and rails, galvanized, and then delivered to the ship for setting up. The time of delivery at the ship is the basis on which the estimating must be done. All parts which go to make up the ship are pushed along in time to be at the ship when it is their turn to be put in place.

Each of these different trades is separate from the other and has an entirely separate organization and the division of work is always sharply defined. There is one General Foreman in entire charge of a department and he has foremen under him on each boat which is under construction. These foremen in turn may have sub-foremen or " quartermen " in charge of gangs.

Preparatory to being used on a vessel, the various materials must undergo a process of "fabrication" which takes it through the Plate and Angle Shop where it is made to conform to the various requirements as laid out in the Mold Loft following the plans furnished by the Hull Drafting Room.

Duties of the Loftsman

A thorough knowledge of ship construction is one of the prerequisites to the work of a Loftsman, and as he is concerned particularly with Lines, he is often called a "Linesman" but he must also be an all-round man on templates and other work. Each Loftsman works with one or two helpers. The Loftsman, having studied the plan of the ship and the table of "offsets" proceeds to lay off on the floor of the Mold Loft the heights and half widths of all decks together with the water lines of the underwater body. The curves are drawn by laying long wooden battens with one edge touching the points which have been set off and the battens secured to the floor by small sharp nails driven alongside. A heavy pencil line is drawn or a light line cut into the floor with a scrieve knife. As the average ship is too long to lay down continuously it is customary to superimpose one part over the other, having the forward part together and the after part on the opposite side of the floor, with the base line parallel to the opposite wall.

Cross-sections of the ship at every frame space are then laid off at one end of the building, and frame lines are scrieved in with the knife and any irregularity in the fairness of the curves, being easily discovered by the loftsman, are entered on the Table of Offsets for the future use of the Hull Drafting Room. While this work is progressing, two other loftsmen lay out templates for floors; frames, side and cant; longitudinals; stringers; hold stanchions; deck beams; brackets; clips and any other angles.

Another loftsman lays off all decks and inner bottom. The position of all plates is noted and templates made, giving the exact size and location of rivet holes.

Two other loftsmen are meanwhile working on the shell plating. This work is generally done for all plates on "skin strakes" except those having special shapes. The "outer strakes" are "lifted" from the actual position on the ship, templets being made by shipfitters. By so doing, all rivet holes and irregularities which may occur are accommodated on the template, and plates thus fashioned make a fair and smooth surface.

Often four loftsmen lay out all bulkheads, detailing each plate, the location of all stiffeners, boundary bars, backing bars, splice angles, staples for continuous stringers, collars, etc. According to the amount of miscellaneous work, there are from one to six loftsmen with helpers, working on templets for foundations of machinery, boilers, fresh-water tanks and all auxiliaries; masts and other miscellaneous trunks and ducts.

Molds, templets (temporary plates) or templates (different names for the same thing) are generally made of either heavy paper or cardboard which will not tear easily, or of wood about one-eighth inch thick and from four to six inches wide. The paper is cut along the edges to suit the size of the plate and the rivet hole pricked through with a rough circle marked in pencil to attract attention. The wood templets are built up and tacked together, the shape being whittled to suit, and the rivet holes bored out with a small auger. Wood templates, though harder to make than others, are more reliable, as they are not so easily damaged nor affected so much by the weather.

All templates are marked before leaving the Mold Loft with the symbol corresponding to that shown on the plan, thus locating their position on the ship and with the size of all holes and notes whether they are to be countersunk or not.

Duties of the Men in the Plate and Angle Shop

The **Anglesmith** works with two helpers, a "hammer man" and a "backhandler." It is his work to bend and weld water-tight staples, boundary bars, door frames and other miscellaneous work from molds and templets from the Mold Loft or shipfitter who has lifted them from the ship.

The work of the **Furnace Man** or **Frame Bender** is to heat in his furnace heavy structural shapes and bend them on his slab to suit templets furnished him from the Mold Loft or shipfitter.

As described more fully in the chapter on "Frames"

the wood templets are laid on the Bending Slab and their outline marked in chalk on the slab. The templet is then removed and the "set iron," a flat bar about one-half inch thick and two inches wide, is heated and bent to suit the chalk line. This is then secured to the slab by means of bent "dogs" (bent, round iron bars) and the various angle and channel bars are formed according to the different set irons used.

The **Plate Furnace Man** works with a furnace which is not so long but is wider than the one used by the Frame Bender. In this furnace are shaped plates which have a sudden change in shape as those at the ends of the flat keel, boss plates around the swell in the plating due to the shape of the boss where the propeller shaft comes out through the hull of the ship, tank margins, some bilge plating of the shell, etc.

For this work he must make a cradle or frame of angle or other plate. The material is heated in the furnace, drawn out and bent into the shape given by the prepared form. It is necessary to finish the bending after the plate has cooled a little as it will become distorted during the cooling and will tend to curl and leave the form it was bent into.

The **Machine Operators** in the Plate and Angle Shop vary in number according to the size of the shop and the volume of work in the ship yard. Arrangements of the these shops vary with the requirements and with the ideas of the men in charge.

The **Mangle Operator** has charge of the rolls which are used to straighten out plates which come into the shop bent, due to an accident. These are rolled while cold.

Other men work on the "Rotary Shears" to cut curved plates, the "Straight Shears," which make straight cuts, and the "Angle Shears," which are used for cutting angle bars.

The **Manhole Punch Operator** punches out standard sized manholes, such as 12″×15″ and 15″×18″ in the clear opening. These are oval holes which can be punched in cold plates of about three-eighths of an inch thickness, as in Tank Top plating.

The **Bending Rolls Operator** has charge of the rolls for shaping plates with a gradual curve which do not need to be heated.

The **Hydraulic Joggling Machine Operator** handles the plates which are to be used on "outside" strakes of shell and decks. The two wheels put a crimp in the edge landing of the plate, of an offset equal to the thickness of the other plates. (Some shipyards do not use this system but rather the "in-and-out" method with parallel liners under the outside strakes.)

The **Planer Operators** have two types of machines. One for use when planing the edge of plates and the other for planing the toes of angle bars (preparatory to caulking).

The **Friction Saw Operator** works in the Angle Shop cutting off angles to the correct length.

The **Punch Operators** are located in both of the shops operating the punches for making rivet holes in plates and angles. These punches vary in design but the modern types, for plates, have adjustable tables so that one man can operate the punch and move the plate to any position required. For the angle punches helpers are still required to handle the bars back and forth into position. The punches for the plates are all vertical, but many are of horizontal action for the angles.

The **Drill Press Operators** are employed with a portable motor suspended from a swinging, horizontal track, which operates either drills or countersinks.

These are used, for drilling special holes and for countersinking punched holes for both angle bars and "outside" plates. The bar or plate is supported by a rack and the motor swings directly over the work.

The **Acetylene Torch Operator** is a part of the Plate Shop working force, as he is often called upon to burn out large, irregular openings which would otherwise require much time to punch or cut out. (Before the days of the torch it was customary to cut out large holes by punching small, say three-quarter inch holes touching each other and then trimming up the uneven edge by chipping.) This man works alone, but all the other operators have helpers, as required.

The material might now be considered "fabricated" and ready for assembly on the shipways. Work of the different gangs employed in putting it together and fastening it aferwards will now be taken up in as nearly the order of the work as possible.

Duties of the "Erector" or "Plate Hanger"

This is the man in shipbuilding practice who places the various parts of the ship in their proper positions, after they have been fitted up or fabricated in the shop.

The Erector has working with him a number of Bolters and Carpenters who carry on certain parts of the work not conveniently looked after by the Erector himself.

The first job which the Erector has to do is to place the keel in position on the blocks. In doing this work he has the assistance of the yard Riggers and a few Bolters who see that the plates are properly bolted up. The various plates are properly marked according to a particular system, which enables the Erector to put each one in its proper place, as indicated on the drawing with which he is supplied.

The flat keel plates when deposited on the blocks are trued lengthwise, to the center line of the ship laid on the keel blocks. The center line of the keel is snapped on the keel plates with a chalk line, located from the holes already punched in the plates.

When laying the flat keels and keelson it is necessary for the erector to be sure that his "overall" length is coming correctly. The distance between the Collision Bulkhead at the forward end of the Inner Bottom and the frame farthest aft that it is convenient to measure to, should be carefully checked. If it is not correct to a fraction of an inch it should be made so at once, by "regulating" the plating, either taking up a little or letting out a little in the length.

It is very important that the distance from the bulkhead at the forward end of the Engine Room to the After Peak Bulkhead should be accurate, as otherwise the length of the shafting as ordered from the drawings will not be the proper length to fit to advantage without too much "packing" between the joints (where the different lengths come together) or else it may be too long, which is worse.

The garboard strakes on port and starboard sides are next placed. Be sure to get the proper lines of rivet holes fair in placing all strakes. It is not necessary to set all keel plates before the garboard and bottom strakes are placed; place a hundred feet of keel, and while continuing work on the keel, more workmen will start the garboard and other bottom strakes, placing alternate port and starboard bottom strakes.

After the bottom of the ship is laid the **Regulators** fair the plates, then the center vertical keelson is placed. This sometimes comes from the shop with the angles attached, and the holes punched in them. The holes in these parts must match, and if the keel has not been properly placed, the keelson will not set true fore and aft.

After the keelson is set, the floor plates may be erected. The Erector must be sure that the proper holes are used in bolting up the parts, or the plates will not stand square with the keel. Midship floors are put in place first, so work may proceed forward and aft at the same time.

As each floor is bolted to the vertical keelson, be sure the proper "liner" is in place on the shell, for the outer strake plate. The shell should be bolted, starting at the garboard strake, as this serves as a guide in fairing other strakes, because it is attached to the keel, and this in turn is attached to the heavy vertical keelson, which holds all parts true.

The Tank Top is generally a level surface, so the floors

are shored up or down by means of shores and wedges, until they are level athwartships. If slight errors have occurred in the manufacture of the parts, the fairness of the shell should be given preference over the tank top. After the floors are erected and faired, they may be bolted solidly, placing a bolt in every third hole. The intercostal longitudinals may now be set.

As the various parts go up, some men are assigned to "lining" and "packing" work. This work consists of putting "taper" liners at the plate landings of the shell, if the plates are not "scarphed," the "parallel" liners between the frame and the outer strake, and as the vessel progresses, of putting liners under the side plates, Tank Top and decks.

As it is customary for all steel to be red-leaded on the surfaces which touch each other, to prevent corrosion due to moisture between the metals, Erectors are required to red lead any of the steel which comes to the ship without having been thus painted.

Stop waters are also placed at water-tight bulkheads and such other places as the design of the vessel may demand. This work is not looked after by erectors, but by another class of workmen. It is the duty of these men, known as "linermen" and "packers," to see that these details are attended to as the ship advances. After the floors are leveled and the shell faired the tank "rider" and "margin" plates are put in place.

This serves to fair and hold the ends of the floors in the proper position, and is a base for placing the side frames. It is not necessary that any one of the steps described shall be carried out over the whole bottom before the next is taken up. The floors amidships may be in place and bolted up, with tank margin plate being bolted while the keel may be going down forward or aft.

The Erector must always be careful to keep his keel true to the line on the blocks, and all holes fairing properly, in order that the ship may be erected true to her designed form.

After the start is made with the margin plate and it is bolted down on port and starboard side by the Erectors, a gang may start on the "bottom shell" plating reaming the sections that are bolted up. This work is done by the Drillers and Reamers, who ream every hole in which there is not a bolt. A gang of Bolters works with the Reamers, transferring the bolts from the unreamed to the reamed holes, the Reamers going over the berth a second time and reaming the holes from which the bolts were taken.

During this time, Erectors have continued running up the side frames and shell, and started laying the tank top plating. Riveters have started work on all berths that have been prepared by the Reamers, and they proceed with riveting up the bottom shell. The side shell should not be finally bolted to the frames, till the frames are set up, the deck beams placed, and the "stringer plate" set. The stringer plate on each deck serves to fair the frames. As the deck beams are placed, their midpoints should be checked up directly over the keel and beam stanchions, and then set and bolted.

As the side frames and deck beams are going up, the transverse bulkheads are set true to the keel, and the deck stringer plates bolted up. After the stringer plates are set, and the frames fair, the shell plating may be faired and bolted. Deck plating may be placed and reaming done on all parts, as already described. The riveters drive the class of rivets required by the specifications.

The shell angles or lugs for various deck stringer plates may be fitted and riveted as the hull goes up.

In the case of the side stringer bars, many of them are curved. If the curvature is slight, the bars may be sufficiently flexible to be pulled into position without preliminary bending; if considerable, or if the bars are of rigid section, each one is bent to shape in the shop. Intercostal shell bars are fitted in short lengths between each frame. These shell chocks (short angles) for stringers are placed above the plate, for they are then more easily fitted and riveted; but if, when so placed, they would foul a shell landing, they are fitted below it.

Special bars are used for making bosom pieces (butt-straps on angle bars) having the same thickness as those bars they are intended to connect, but flanges about half-inch narrower. To secure proper caulking in bosom pieces of water-tight work, the rounded toe should be removed by planing or chipping. The edge should be sufficiently sharp and square for caulking purposes. The ends of joined bars should be smoothly cut.

The tank margin plates amidships are straight fore-and-aft, but those towards the ends have curved outer edges. In placing deck plating the stringer plates are often the first dealt with, for when they are in place they serve to hold the frames fair and at the proper distance. The position of the various plate landings is determined by the holes in the plate and on the beams. A working plan is provided for each deck, showing every feature affecting the plating, and giving figured dimensions for all distances and sizes. When the weather deck stringer plates are up in place, the stringer angle may be fitted.

The necessary information for placing shell plating is provided on an expansion plan of the shell. The position, form, and thickness of all doubling plates and the size and positions of cargo ports, side lights and scuppers, all water-tight bulkheads, tank margins and divisions; the decks, also intercostal stringers, keelsons, bilge keels and fenders are shown; also the breadths of the landings, lap joints and buttstraps, size and spacing of the rivets. The different classes of riveting may be shown by marking across them one, two, three, or four short lines. Large-scale sketches are provided to illustrate the method of fitting special parts, such as the disposition of the buttstraps, the rivets in the shear strake joints, and the tack rivets in the doubling; the mounting of cargo ports, etc. It is important that every feature affecting the fitting of the shell plating should be clearly understood, so that no trouble arises as the ship is built.

The parts of a cellular double bottom are differently arranged at various localities; under the engines, for instance, additional intercostals are introduced, and towards the ends the ordinary line of intercostals may step inwards towards the center line. These variations are indicated to the workman on the plans and frame list.

In shipyards which have facilities for handling heavy weights on the shipway cranes, it is possible to build many of the bulkheads on the ground, in front of the ship, and then raise them into position after they have been riveted. When a bulkhead is too heavy to handle at one time, it is assembled and partly riveted but bolted through a line which is left loose and then lifted in two or three parts. The balance of the bulkhead can be riveted and after all the pieces are in position on board the ship the parts which were left loose are riveted and the whole job is ready for caulking, provided the bulkhead is one which is to be caulked.

When assemblying any such work on the ground it is necessary to be sure that the overall dimensions are right. The width and height should be tested with a steel tape before the work is bolted so that it will surely fit when it has been lowered into position on the ship.

During the erection of the ship, measurements should be taken from time to time to ascertain if the width is correct and if the heights of the beams along the center line of the ship are correct. Too much care cannot be exercised in the early stages of the construction work because it is then that mistakes if discovered can be most easily remedied.

In some of the ship-yards it is customary to have a gang of men called "Regulators" whose duty it is to go carefully over all the plating and be sure the Erectors have laid it with "fair" rivet holes. If they are not fair then the Regulators make slight adjustments, a little sideways movement on adjacent plates, which brings the rivet holes more nearly in line.

Some of the ship yards do not have a separate gang of

men for this work but have the Erectors fair the material as it is being placed.

Notes on Erecting

The tools used by the Erector (described in Chapter III) are simple and few. It is well to wear a broad leather belt with a pad and hole in it for carrying the "spud" wrench when both hands are needed while climbing about.

Sometimes a plate is to be swung sideways and the workman is ahead of it so he sticks the handle of the wrench into a rivet hole and having a firm grip on the plate, can pull it toward him and swing it around as he desires.

When a plate or angle is nearly in position, the workman will often fit the handle of his wrench through a rivet hole and catch the rivet hole which is to come directly under the one which holds his wrench, with the point of it. In this way he is able to center the hole and when the plate is finally lowered it will fall directly into place, guided by the handle of the wrench.

The Erectors are classified as "cranemen" or "hand workers." The former are usually men of experience and use a crane or derrick for handling heavy loads, while the "hand gang" are often laborers under an efficient leader who handle brackets, braces and other small parts which can be lifted by a few men and fitted into place.

The Erector, probably more than any other operator on a ship, must keep in mind that upon his care and diligence depends the safety of nearly every workman on the ship. He must see to it that the plate he is carrying at the end of his tackle is securely fastened and that it is not too heavy for the capacity of the boom. In his work common sense, a few facts and a little caution are the greatest factors.

The derrick falls (tackle) generally have a large hook on the lower block. From this a "sling" of either chain or steel wire rope is used to hitch onto the material. The chain has a large ring for the fall hook and smaller rings at the lower ends, which are fitted with shackles and bolts with nuts. The wire rope sling has a large ring with two parts leading from it having rings and shackle similar to the chain sling.

As a general rule shackle bolts of $\frac{3}{4}''$ will safely carry loads on a sling up to two tons while one-inch bolts should be used for heavier loads.

Plates are lifted either flat or on edge. When lifted on edge the bolts in the shackles on the sling are fitted into rivet holes near the upper edge of the plate at about one-fourth of the length from each end. For lifting a plate flat it is necessary to use "lugs." These are usually $5'' \times 5''$ angle bars about $\frac{3}{4}''$ thick. One flange has a hole about $1\frac{1}{4}''$ diameter in the middle of the length; the other flange has three oblong holes about $1''$ across and $2''$ long. Two of these lugs are bolted to the plate, near both ends, as far as the sling will spread easily and the shackle is bolted through the one hole in the standing flange.

The following are a few safety measures that it is well to keep in mind: Always fit a washer under the nut when bolting a lug to a plate so that by no chance will the nut pull through the rivet hole: Always put a nut on the end of the shackle bolt when raising plates, angle bars and other shapes on edge so that in case the material being raised knocks against any other part of the ship the bolt will not loosen and drop out, letting the end fall and possibly breaking the other part of the sling: A "togline" (a light weight rope with a hook on one end) in an open rivet hole in the material is an easy way of keeping a plate from swinging and for guiding it into place: Be sure that the falls of the derrick hang plumb over the material to be lifted and that there is no twist in them, as it is quite difficult to control a load swinging as the result of this untwisting.

The signals opposite are easily understood when directing the operator of the derrick boom or crane in handling the load. Make motions with the hands slowly and distinctly so that they may be quickly seen and followed.

Duties of the Bolter

The work of the Bolter is to bolt the parts of a vessel solidly together after they have been "regulated." He may assist the Erectors or "Linermen" in placing the various liners. The purpose of the liners mentioned will be explained later.

The Bolters need an ample supply of bolts, nuts, washers, and liners. Some bolts are $\frac{1}{8}$ in. smaller in diameter than the holes in the plates, and some are $\frac{1}{4}$ in. smaller. This latter size bolt is necessary for bolting certain plates in which the holes come unfair, but in no case should the bolt be allowed to stand other than square with the floor; that is, never try to "set up" a bolt when it is passing in a slanting direction through the plate.

A bolt should never be forced through an opening, as

Stop
Boom Around
Load Up
Load Down
Boom Up
Boom Down
Free Hand Near Signaling Hand Move Slowly
Boom Down
Load Down
(Both Hands)
Boom Up
Load Up
(Both Hands)
Load Down
Boom Up
(Both Hands)
Boom Down
Load Up
(Both Hands)

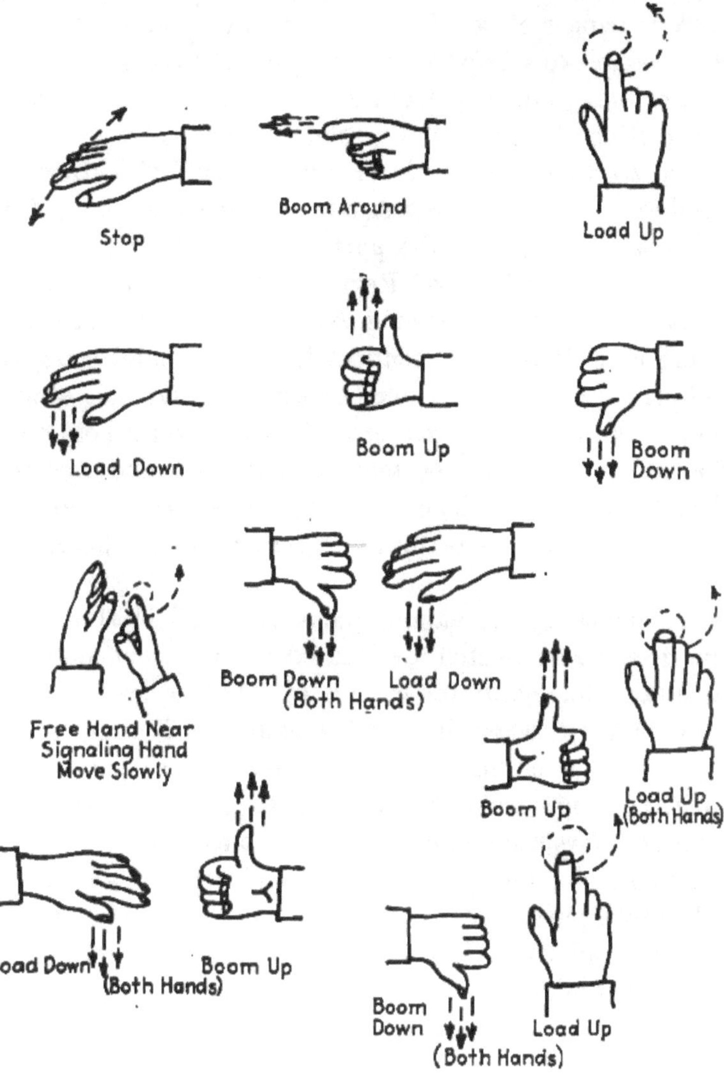

FIG. 1

there is danger of bending over the threads and rendering it impossible to set up the nut as tightly as necessary.

For lining up the plates the Bolter also needs a "drift" pin, a sledge hammer, and a solid wrench.

In starting a ship the Erectors have with them a number of Bolters, who assist in erecting and put a few bolts in place to hold the parts in the desired positions. They also work with the "Regulator" in lining the various parts of the ship. Following the Regulators and the small squad of Bolters, just mentioned, comes a large gang of Bolters who line up all holes by drifting, and bolt everything solidly for the Reamers. As the Riveter goes over the hull in his work, he removes the bolts and replaces them with rivets. The rivets placed in the holes not bolted hold the parts firmly together as the bolts are removed.

In punching the various parts of a hull in the shops the holes to be riveted are located in the exact position necessary for accurate assembling; but a great many come "unfair" when the ship is erected and if after "drifting" a number of "half blind" holes are so close together as to make it impossible to fit the bolt properly then these "unfair" holes should be reamed out so as to get in a good sized bolt with which to draw up the plates good and tight. A length of pipe fitted over the handle of the "spud" wrench will give additional leverage. The longer the pipe the easier the work of tightening the bolt becomes and on a deck or tank top where the plates are horizontal the pipe may be three or four feet long.

Whenever a countersunk hole is enlarged to correct

nfairness, the countersink should also be enlarged to ccommodate the larger rivet. Small rivets must not ›e used for riveting up unfair holes.

The Bolters, in clamping up the parts, must see that he irregularities are removed by using a large number of ›olts to clamp the plates in place against the frame. Vhen plates are bolted up they must look true and even.

Liners are laid as required, and must be well fitted, ›y means of a template giving the required lay-out of ıoles. These liners are made in the smith shop and ›unched before placing.

The Bolter, in addition to bolting frames, shell, and ecks, bolts up the bulkheads. Plates are located in the ame manner as described for the shell and deck plating nd this work presents no new features.

In setting up his work the Bolter must be sure it is ove up solidly, so the rivet may not spread out between he plates during the staving up process. If holes are rilled in place in such position that they cannot be olidly bolted, the Bolter should follow after the Driller, nd slack off such drilled plates, cleaning out all "chips" nd "rags" that may have lodged between the faying sur-aces. The "faying" surface is that surface of the late which the punch enters. (On its exit it leaves a ' burr.")

Many times it is found to be impossible to draw up the naterial as should be done without first heating it and using he "flatter" and "maul." The average length of a bolt on ork of this kind should not require much "packing" under he nut, but when more than three or four thicknesses of scrap iron or washers are fitted, a different length bolt should be used.

In bolting up a section about to be riveted a bolt should be placed in every second hole, for heavy work. If these bolts are well hove up, the joints will be solid and the riveter can do good work very rapidly. If bolts are placed too far apart, the rivet swells out between the two surfaces of the plates being riveted and a tight joint is not possible. Very close bolting is necessary, too, because the large amount of hammering which a long plate receives in riveting will lengthen it out and make a great many unfair holes if the bolts are not placed close enough to prevent movement. However, Bolters must avoid the tendency of putting in more bolts than necessary and should work to a standard, using an established number of bolts according to the location and condition of the plates.

While the Reamers are working in a "berth" a couple of Bolters follow along and move the bolts, heaving them up solidly. The bolt first placed is removed, to a new hole, and the Reamers go over the berth again, reaming all holes in which bolts were first located. After the Reamers have completed a berth it is ready for the Riveters.

The Bolter must be careful in his work to see that the holes are true, and in line, when leaving work for the reamer.

He will find it impossible to make all holes stand perfectly true, but by the use of the drift pin he must get the greatest possible number into line, having no "half blind" holes.

(A "berth" is a space arbitrarily fixed by the fore-nan riveter between certain frames, fore and aft and from ıne side of the ship to the other.)

Duties of the Drillers and Reamers

The duty of the Driller and Reamer in shipbuilding s to drill, ream, and countersink various parts of the hip which must be machined after they are in place ın the frame. Drilling consists of cutting holes in the rarious parts at selected positions by the use of small ools known as drills. Reaming is finishing out holes that nay be unfair and countersinking is enlarging the face of ither drilled, reamed or punched holes.

Compressed air for operating the air motors, riveting, hipping and caulking hammers is carried from the Air Compressor House by means of steel pipes which are led long the shipways, some beside the keel and others are run n "risers" up the side of the ship, near the staging, and then ınto the deck of the vessel.

"Manifolds," which are cast-iron fittings with a number ıf outlets, are screwed on the end of these pipes. In this vay it is possible for six or more different lines of hose to be ıttached at one manifold, as the connections at the manifold ıre made to fit the hose couplings.

The majority of shipyards use rubber hose which is nanufactured to stand a heavy air pressure without bursting. A few yards use hose which is "armored," protected ıy wire.

At the ends of the hose a patent coupling is attached. This is made up of a brass fitting which fits tightly and can ıe connected by pressing it against another fitting, giving a sharp turn and it slips into place, a guard on the fitting holding it in position.

Drilling

The air motors are made up of a body with two handles, one being the dead handle, and through the other the air flows to the interior, through a hose attached to the end. On the upper end there is a feed spindle operated by a handle which is used to maintain the pressure through the machine and down to the point of the drill. The lower end is in the form of a chuck which will take any tool having a standard Morse taper shank from $\frac{7}{8}$ of an inch up to about 2 inches.

This chuck is made to revolve by means of a gear wheel at its upper end, inside the body of the motor and the speed of revolution is increased by means of other gear wheels. The air as it passes in through the handle is circulated around to separate pistons. These are set so that they avoid a "dead center" and are all connected to the same crank shaft, each having a separate connecting rod. The crank shaft has on its end a geared wheel and it is through this that the power of the expanding air pressing against the pistons sets them in motion and thus revolves the crank shaft; turning of the gear wheels results and the chuck revolves at a high rate of speed.

The drills and reaming tools are held in this chuck and in this way the motion is carried down to the end of the tool. Lubrication of these moving parts is very necessary and a thick heavy oil is plentifully supplied to reduce any friction which may be caused as the speed of these pistons is rapid. This lubrication should be done in the tool house rather than by any attempt of the men on the ways. The body of the machine is constructed dust tight and the parts when once well lubricated remain in working condition for a considerable time.

Sometimes through an oversight in the Tool House some of these motors are not lubricated, as often as they should be. This can be detected by the operator noticing a jumpy, jerky motion and a rather sluggish response when the air valve is turned on, as there is a tendency for the moving parts to bind when not well oiled.

Sometimes the parts have been lubricated to excess, and when first using the motor some of this oil will pass out through the discharge port with the air. This will continue until the extra oil has passed off and the operator should not take any further notice of it as it will regulate itself automatically.

This description of the working parts and lubrication applies equally well to the "round machine" and the "corner machine." The shape of the two machines or motors is dissimilar, but the design of the interior of the working parts is very much alike. Some of the operators are accustomed to using a hose connection on the handle but it is oftentimes desirable to use a screw connection on the handle with a short length of "leader hose" (about 36 inches). This leader hose should have a screwed connection to the handle and a hose coupling at the other end. This is done to avoid a tendency on the part of the patent coupling to jump off the handle due to the fact that the handle is absolutely rigid. With this flexible length of leader hose, this tendency is done away with.

Any drill under $\frac{7}{8}$ inch requires an intermediate socket. If the drill is $\frac{7}{8}$ inch or over, be sure the shank enters the chuck properly, after first screwing back the ejecting pin at the top end of the motor to allow the drill to come up into place.

Before drilling it is always necessary to lay off the center of the intended hole and mark its location with a deep impression formed by a "center punch," and where accuracy is expected, four "tell tale" punch marks thus . : . are placed a little outside of limits of hole to be drilled for checking after hole is drilled and the original center mark is obliterated.

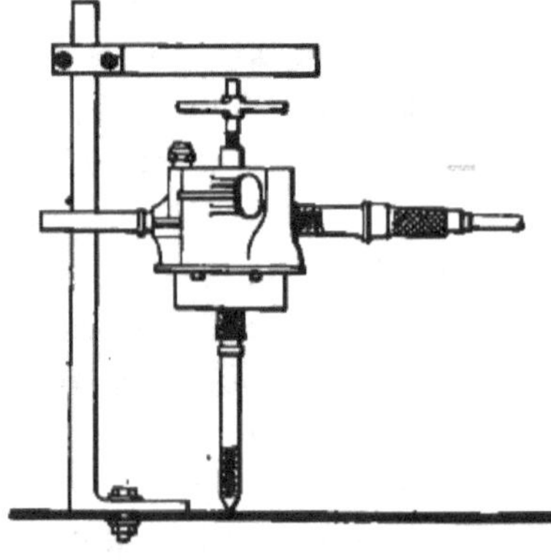

FIG. 2.

The shank of the drill fits easily into the chuck and should never be hammered, in an effort to force it into place, as hammering will damage the drill.

When using an "old man" on a horizontal plate (Fig. 2), bolt up the base of the "old man" so that the arm can swing around over the location of the hole, taking care if possible to have the base in line with the arm. As soon as the hole has been located with a center punch, having the drill in the air motor, put the point of the drill in the center punch mark and place the movable arm of the "old man" against the "feed wheel" of the motor, taking care that this feed wheel has been run down into the machine before the drill is adjusted. Then clamp the arm of the "old man" in place.

Now lubricate the point of the drill so the oil stands on the plate, but do not put any oil to the point of the drill until the full size portion of drill enters the plate, as oil obscures view of the hole when starting. Before turning on the air swing the motor around so that the dead handle will come against the upright of the "old man," being on the left hand. Facing the machine, with the right hand on the throttle turn on the air by a slight movement of the wrist, holding the feed wheel with the left hand.

Advance the speed of the drill by giving a little more air. Increase the speed to normal and lubricate the point from time to time as it becomes necessary to prevent it from smoking due to the heat created by the friction of the dry steel, taking care at all times to keep the pressure on the point of the drill by screwing the feed handle up toward the arm of the "old man." It is necessary to have a uniform, regular feed, so that the progress of the drill is steady.

The driller should be careful to observe the advance of his drill into the material and when it appears to be near the bottom of the plate he should be on the alert to notice a jumping movement of the machine as the drill works through the under surface from the small hole which it makes as the point of the drill first goes through, up to the larger diameter which would finally be cut, equal to the diameter of the drill. The drill will give a series of sharp, quick jumps as the feed is not now given as quickly as the advance of the drill and as it breaks through the under side of the plate, the feed wheel will fall out from under the arm of the "old man," against which it has been pressing.

At this juncture in the operation the workman should shut off his air, and the speed and care with which he works will give either a clean cut through the plate or maybe a broken drill. When this jumping motion is detected just as the drill is going through the plate, the workman should grasp the dead handle with his left hand to steady the motor, and turn off the air with his right hand. He should steady the machine so that it will not tend to fall over on its side but will remain upright with the drill down through the hole. In order to withdraw the drill from the hole, run the feed wheel back into the air motor to give room above the motor and then lift the air motor and drill directly up in a vertical line, providing the hole has been cut clean.

If there are any burrs around the cut it will be necessary to turn on the air again, holding the machine as rigidly as possible and with a few revolutions of the drill cut off any burrs remaining.

It is necessary that the drill should be normal to the plate and in order to be sure of this it is a good plan to have a block of wood which has an end and one side at exactly 90 degrees and use it about the same as a carpenter's steel square is used. Place the end of the block on the plate and swing the motor until the drill touches the vertical side, evenly along its length. Then swing the block at 90 degrees to its first position and be sure that the drill again touches the vertical side along its length. In this way the operator can be sure that the drill is perfectly normal to the surface of the plate.

Due to an insufficient depth of the center punch mark or the poor grinding of the point of the drill occasionally when it starts the drill will walk off sideways. This may be overcome by swinging the head of the motor in the same direction, thus tending to start the drill back toward the center of the hole.

When it is desired to drill on vertical plates or on overhead work the same procedure should be followed. It is a

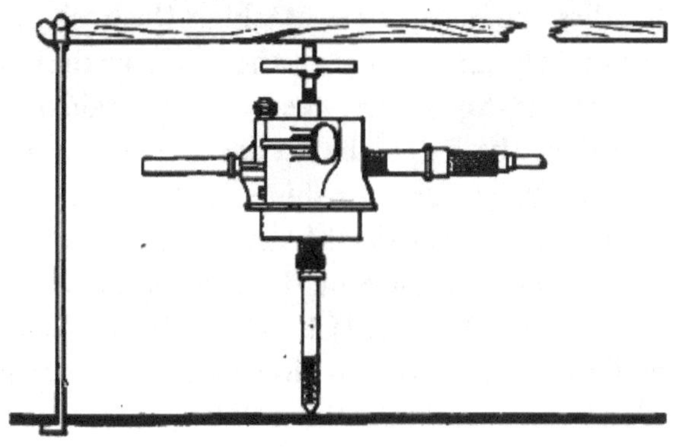

Fig. 3.

matter of "safety first" to wear goggles over the eyes at all times, but for any of the overhead work it becomes imperative, as otherwise the chips are very apt to fall into the eyes of the operator.

When the driller has an extra man with him as a helper it is often possible to use the "hook stick" in place of the "old man," but it can be used *only* when there is another man with the driller,—(Fig. 3). The hook stick is simply a piece of joist about $2'' \times 4''$ and about 6 ft. long. At one end a hook is secured to it by means of an eye bolt, the lower end of this hook is turned over at right angles so that it can be easily slipped down into a rivet hole and caught on the further side of the plate, the length not being much over $1\frac{1}{2}''$. The hook is usually made of about $\frac{5}{8}''$ round iron and the length is equal to the combined length of the air motor and drill (about 26 inches over all).

This hook stick is operated by the helper. The drill and motor are placed in position by the driller, and his helper handles the hook stick, catching the hook through a nearby rivet hole and swinging the stick so that it comes directly over the top of the feed wheel, against which it presses. When the helper has it placed right and holding it firmly, the driller turns on his air, steadying the motor with his left hand on the dead handle and his right hand on the throttle. The advance of the drill is taken up by the helper, who lowers his hands to suit. In this position the dead handle is pushed toward the operator, but he can hold it off with a "stiff arm." As the drill begins to work through the under side of the plate a warning is given by the action of the air motor, the helper eases up the pressure on the hook stick and as the drill goes through the plate he removes the hook stick from on top the feed wheel. The motor is then withdrawn by the driller as previously described. The hook stick is used generally for vertical work where the drilling is downward.

When it becomes necessary to drill up through the bottom side of a shell plate the thrust of the drill can be taken by building up a series of blocks from the top of the shipway, allowing room to fit the motor and drill into position, and resting the feed wheel against the upper surface of the top block; a deep center punch mark should be made on the side of a drift key and this should be used under the point of the feed spindle for it to press on.

For work on the outside of the "side shell" a support for the drill can be rigged up by placing a heavy plank or joist against the shipway staging and resting the feed wheel against this. In case the distance between the ship and the staging along the side is too great, the extra distance may be taken up by adding an extension between the drill and the motor, thus placing the drill against the shell of the ship.

When drilling through bulkheads no set rule can be given for rigging up any apparatus, as this would vary according to the location of the bulkhead and the conditions at the time; but generally speaking if the hole is to be drilled on the side on which the stiffeners are located it will be possible to use a hook stick, catching the hook under the flange of the channel bar stiffener which is away from the bulkhead. If the hole is to be drilled on the caulking side of the bulkhead it is usually possible to rig up some backing for the feed wheel, as the riveting is done from this side, and uprights for staging for the use of the riveters are generally in position for quite a while during the construction of the boat.

There are many cases where it is impossible to use a No. 3 "round" motor (which is used for general work) as for instance when holes are to be drilled near engines, boiler foundations, or other places in the Engine and Boiler

Rooms, where the space is very cramped. It then becomes necessary to use the "corner motor" in which the operation of handling the drill is the same as for the other type but the feed is governed by a ratchet wrench instead of the feed wheel. The reaction of the machine against the turn of the drill can be taken up by swinging the machine around until it touches a portion of the ship structure which will then keep it from turning. This must be done before the air is turned on. The air valve and hose connections are the same as before. The corner machine should be used only in those spaces where it is impossible to use the round motor because it is unwieldy and unbalanced and is fitted for use only in places which are inaccessible to the other type of motor.

When drilling odd work, such as butt straps to the shell, where the inside and outside straps are fitted with punched holes the plate is blank, it is well to drill a small-sized hole for a pilot and then ream it to the required size. This allows an opportunity for striking centers with the holes which would be difficult in case a large-sized hole was drilled out first.

When poor riveting has been done in watertight work and the rivets must be removed so that new rivets can be driven in their places it is customary to drill material from the point of the rivet so that it is possible to back it out. It is necessary to drill the hole down through the point into the rivet until the end of the drill has passed beyond the countersink and down into the body, between $\frac{1}{8}$ and $\frac{1}{4}$ of an inch. After this has been done the rivet is forced out of the hole head first by means of the "backing out punch."

Taking out the metal in the middle of the rivet allows the bevel sides of the point to crowd into the space thus formed and the flare of the rivet point is thus able to pass through the hole for the body of the rivet. It often happens that by drilling through the rivet in this way it reduces the metal so much that when the workman starts to back out the rivet it breaks off and leaves the point remaining in the countersink. This is easily removed and the hole is ready for another rivet.

Tapping

Many times it is necessary to drill a hole in the material where it is not going all the way through. It often happens that rivets cannot be used, due to local conditions and a "tap rivet" is used in its place. A tap rivet is a threaded bolt with the head formed for countersunk work, and has a small projection beyond the head for attaching the wrench used to turn the tap down into place.

These tap rivets are often used when fastening plates to a heavy casting, where the thickness of the casting will not allow through rivets. The hole for this work is drilled in the manner previously described after which it is ready for the "tap," if it has been cleaned and the outside surface of the plate is free from any chips, so that there will be no chance for anything to catch as the tap enters the hole.

A tap is a bolt with a head for a wrench at one end and teeth at the other end for cutting the threads to fit the threads of the bolt. These teeth are fluted to give a cutting edge, or entrance to the tap and are in five groups. These groups run up to about half the length of the bolt and are separated by means of the fluted spaces. The teeth are the same as are seen on any bolt and have the same shape and inclination or spiral direction. If the tap is revolved it will advance into the hole in the same way that any bolt will advance when it has been turned. In order to give a small clearance these teeth are a trifle larger than the teeth of the bolt which would be used in the hole. The tap is made of tool steel with a fine cutting edge so that it will cut and force its way through solid steel or other material which forms the wall of the hole that has just been drilled. These taps are made in three numbers for each size of hole. Number 1 being the smallest is the first one used.

By using the Number 1 first, sufficient metal is cut in the walls of the hole to act as a guide and yet this first cut is not so deep that it will require considerable power. Number 2 cut is a little deeper and Number 3 gives the final depth of the thread.

When starting Number 1 tap it is necessary to have it exactly in line with the hole so that the threads as they are cut will be uniform about the center. Whenever possible, due to the nearby structure, it is necessary to grasp this tap on both sides of the handle in order to hold it straight in place. Difficulty will be encountered in corners where this cannot be done.

After the first two threads have started into the material the tap will guide itself along the line of the hole. The depth of the hole is usually drilled to an amount equal to the diameter, i.e., a ½-inch diameter would be drilled to ½-inch and a 1-inch diameter would be drilled to a depth of 1-inch.

In all cases the tap should be carried down to the full

depth of the hole, thus securing all necessary threads for the full strength of the tap rivet or bolt which will be used. When starting the first tap it is important to use plenty of lubricating oil as the cutting will be rendered much easier and any tendency to injure the cutting edges of the thread will be reduced.

When turning these taps it is necessary to revolve them part way around and then turn them in the opposite direction for a short distance, thus forcing the "cuttings" out into the fluted space between the groups of threads and avoiding a choking up due to material. This backward turn need only be slight and is followed by another advance. In other words advance about one-half of a revolution and retreat a quarter turn to clear the material, following this up by an advance of another half revolution where the first had left off.

These taps are operated by means of a wrench which is supplied with them. It is generally a round handle with one flat end through which a square hole has been made which engages the square part of the upper end of the tap. When turning the tap sometimes it is possible to swing this handle all the way around and if two men are working on the hole (they generally face each other), one takes the handle as it comes toward him, passing it around a half turn, when it is taken by the other man. When the hole is being tapped by one man he generally removes the handle from the tap and revolves it half a turn, thus keeping the handle on the half circle between him and the tap.

After the taps have been used it is well to be sure that they are properly cleaned, being careful to go between the teeth with a rag or waste of some kind to remove all chips and oil.

The following table according to the U. S. Standard thread gives a basis for using the necessary size drills and taps to fit the required sized taps.

Tap	Threads Per Inch	Drill	Tap	Threads Per Inch	Drill
1/4"	20	3/16"	7/8"	9	3/4"
5/16	18	1/4	1	8	27/32
3/8	16	5/16	1 1/8	7	31/32
7/16	14	3/8	1 1/4	7	1 3/32
1/2	13	13/32	1 3/8	6	1 7/32
9/16	12	15/32	1 1/2	6	1 5/16
5/8	11 1/2	17/32	1 5/8	5 1/2	1 13/32
11/16	11	19/32	1 3/4	5	1 1/2
3/4	10	5/8	1 7/8	5	1 5/8
13/16	10	11/16	2	4 1/2	1 3/4

Drilling Speeds	High-speed Air Motors
1/2" drill................500 r.p.m.	"Corner Machine"....... 150 r.p.m.
5/8" " ....425 "	No. 2 air motor......... 500 "
3/4" " ....350 "	No. 3 " "........... 700 "
7/8" " ....300 "	Grinder...............3000 "
1" " ....275 "	

It is safe to *start* high-speed steel drills at a peripheral speed of 70 ft. per minute for soft tool and machinery steel, but the speed maintained should be a matter of judgment regulated to suit the hardness of the metal being drilled. High-speed drills work much better when warm, often giving good results when the chips are turned blue by the heat generated. High-speed drills are extremely brittle when cold and in cold weather should be warmed before being used.

LUBRICANTS FOR CUTTING TOOLS

Material	Drilling	Reaming	Tapping
Soft steel..........	Oil	Dry	Oil
Wrought Iron.......	Soda water	"	"
Cast Iron..........	Dry	"	"
Brass..............	"	"	"
Copper............	"	"	"
Babbitt...........	"	"	"

Reaming

The trades of drilling and reaming are often combined in shipyard practice and a man who is able to do one should be able to do the other. Rivet holes of the correct size are punched before the plates and shapes are erected on the ship. If the boat has been well regulated these rivet holes should come into a very close alignment, but as it requires only a very small amount of metal to resist the passing of the rivet through the hole, it is necessary that the passage through the combination of plates or plate and angle should be accurate and of the required diameter. As the hot rivet expands the size of the hole should be $\frac{1}{16}$ in. larger for sufficient clearance.

In order to have this condition right for the riveter it is necessary to ream out the holes after the plates have been erected and regulated.

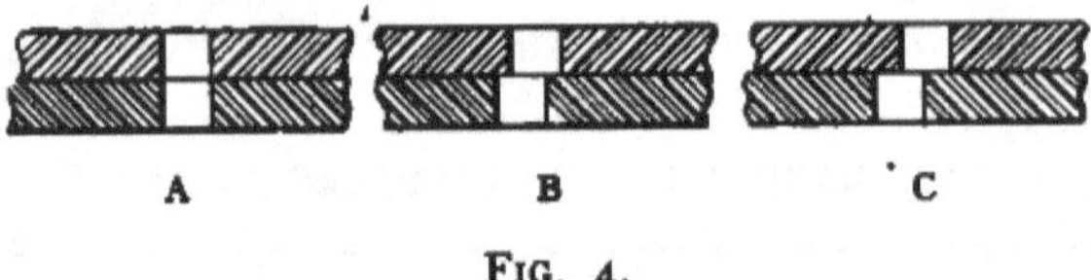

A B C

FIG. 4.

In Fig. 4 at *A* the rivet holes are "fair." At *B* the rivet holes are shown which can be reamed to a larger size. As the hole is about "quarter blind" it would be considered the limit on which the reaming would be allowed, while at *C* the rivet holes are "half blind," a condition that would cause rejection of the whole plate.

After the ship has been bolted up the Reamer reams out the open holes and afterward the Bolter-up changes the bolts to reamed holes and the reamer goes over the work a second time and reams out those holes which before were occupied by bolts.

The air motor used for drilling is also used for reaming. Number 3 air motor will drive reaming tools up to $\frac{13}{16}$ of an inch in diameter. For a larger size hole the Number 2 air motor is required.

Reaming tools are made with fluted sections with considerable recess between each cutting edge forming the flute.

The upper end of the reaming tool is shaped to fit the chuck of the air motor and continues with a parallel side down to nearly $\frac{3}{4}$ of the length of the cutting edges when they taper to a small diameter at the end. In this way it is possible for the reamer to enter a small-sized hole and ream to a larger diameter.

It is necessary in order to make a hole sufficiently large to accommodate the rivet to carry the reaming tool down into the hole in the plate until the workman is sure that the parallel-sided portion is in the hole.

Sharpening of the cutting edges is a very difficult operation for a workman when working by hand, and this should be left entirely to the regular tool grinders in the tool shops, where the proper appliances to give the necessary bevel and slant to the tapered end are at hand.

A gang is often made up of one reamer and a helper, the reamer having charge of the machine, and the helper assisting to take the weight and reduce the labor of the work.

The two men should stand facing in opposite directions, each with the side of the handle of the machine against his left leg, the body of the machine near his right side. The operator should have his left hand on the air valve and his right hand on the feed wheel, the helper facing in the opposite direction, holding the dead handle in front of him. In this way the reaction of the machine due to the turning of the reaming tool is overcome.

When reaming, the machine should be held as nearly square with the plate as possible. In no case should the plate be cut outside of the limit of the edge of the countersink.

It is customary to fit an extension socket between the air motor and the reaming tool, when reaming on deck or a tank top where the work is below the man's feet, thus allowing the workmen to stand erect instead of constantly bending their backs.

When reaming up through the bottom shell with the work overhead, it is necessary that the extension be used again, so that when the reaming tool is forced up through the plate the air motor will be low enough so that the men can get a good grip on it without having to bend the arms. In this way the workmen are able to steady the handle of the machine against the hip and when the thrust is given for the reaming tool, it can be obtained by pulling up on the handle of the machine rather than pushing upward.

On this work the operator of the motor places his right hand on the air valve and guides the reaming tool into the hole with his left hand around the extension socket. The man on the dead handle side keeps the machine from whirling around and assists the operator in handling the weight, but only in such a way as to give the operator full control of it.

For all this overhead work it is necessary that the workman should be provided with safety goggles, as chips and dirt are annoying and often dangerous.

When reaming on the side shell or on a bulkhead the operator should have his left hand on the throttle and his right hand on the feed-screw handle.

For this kind of work it is not necessary to use an extension between the reaming tool and the motor, as the workmen can get up close to the work.

When operating in tight places where the "dead handle" must be taken out, a plug should be screwed into the handle hole to keep the grease in and dirt out of the machine.

When holes are unfair, it is necessary to hold the reaming tool sufficiently "cockeyed," or at an angle with the plate to get the end of the reamer into the little opening, but as the tool cuts into the steel it should be straightened up as much as possible in order that the rivet when finally driven shall be nearly if not quite straight across the plate.

Holes on outside plates are often countersunk in the Plate Shop and when reamed out on the ship they often have not sufficient countersink left around the hole to give a good grip on the plate. To overcome this it is necessary to re-countersink all such holes.

Care should be taken by the workmen to be sure that the tool they are using has not been ground down by re-sharpening so that the diameter to which it will ream is really smaller than it should be. If a tool has been ground off on the edge by resharpening a number of times, it will naturally follow that the hole cut by it will be too small. This condition may not show until it is time for the riveter, but when the riveting gang is not able to force hot rivets up through the hole, it is too late in the game for good workmanship. This matter should be carefully watched by the man in charge of the reaming and should be checked up by the workmen themselves, as they are the ones who will be blamed if the holes are too small.

Many of the shipyards use two tools for the operations of reaming and countersinking, but there are two manufacturers who have on the market a combination reamer and countersink which will perform both of these at one operation. These tools are much in demand by the workmen, but when roughly handled they are very apt to break.

Duties of the Riveter

The rivet, though small, is an important element in ship construction. The proper placing of the rivets in the various parts of the ship is an important work. It requires skill, and the workman should be rapid in doing his work, in order that the building of these ships may not be delayed.

A ship riveting gang is built up a little differently from gangs doing construction work on steel bridges or buildings. In the ship gang each man is usually trained for certain work, the riveter driving rivets all the time, while the holder-on does that kind of work throughout the day. While in construction work the gang is often made up of men who are equally experienced and take turns at holding-on and riveting. On ship work the riveter is considered the mechanic of the gang and he is held responsible for their output.

Whenever the work of the gang is being checked over it is always the riveter who is praised or condemned. For that reason all riveters on ship work should remember that it is only by hard, aggressive action that they will be able to make a big drive and that it is necessary for them to have complete control of the other members of the gang, holding them right down to the work and keeping them all together, as the absence of one destroys the team work of those who remain. The work of the heater boy, getting his fire ready in time and having a plentiful supply of coke and rivets for the day's work, should be watched by the riveter as well as by the holder on.

"Safety First" is quite a consideration with the riveter, as many times he is obliged to work on outside staging on various parts of the hull. Stout shoes that are not slippery on the sole are a big consideration as well as the right kind of hat on his head. This may seem a small matter, but many a time the writer has nearly fallen off the staging because he was wearing a hat of different shape than usual, bumped it against some of the overhead staging and nearly lost his balance by trying to recover himself.

It is necessary to be careful in handling the gun to be sure that the trigger does not catch either against one's finger or clothing, so that the plunger will be shot out and either lost or hit some other workman. Chisels for chipping rivets should be kept with a good sharp edge, as many times work is delayed when these become dull. All the other little details which go to make up success or failure of a day's work should be watched.

Most of the joints used in shipbuilding work are **lap-joints,** where one plate is lapped over another, in order to complete the construction. In certain parts of the ship a lap joint occurs over another plate, and in this case a " liner " must be placed. Ordinarily the Erector or Bolter will see that these liners are in place, but a Riveter should not rivet up a joint until he is sure that it has been properly lined.

In addition to the lap joint, which is the most common in ship work, the **butt joint** is used. Back of the joint there is placed a butt strap which reinforces it and holds the joint firmly together.

Regardless of the kind of joint that may be used in

shipbuilding, a number of different styles of riveting may be introduced. Thus, there may be a single, double, treble or quadruple riveted joint. A single riveted joint is a joint having one line of rivets; double, two lines, treble, three lines, and quadruple, four lines. Except in a few cases, in ship construction the following sizes of rivets are used: $\frac{5}{8}$ in., $\frac{3}{4}$ in., $\frac{7}{8}$ in., 1 in., and $1\frac{1}{8}$ in. Designers make an effort to limit the sizes of rivets to as few as possible. Rivet holes are always punched or reamed larger than the rivet that is to be driven. The finished diameter of the rivet when staved up is larger than before it was riveted. In specifying the size of rivets, it is always the diameter before staving up which is given. The allowance made for the swelling of rivets is roughly taken from $\frac{1}{16}$ to $\frac{3}{32}$ of an inch. The latter dimension is the amount which a $\frac{7}{8}$-in. rivet will swell in staving up, and the former, the amount a $\frac{5}{8}$-in. rivet will swell.

In riveting watertight joints, the rivets are placed closer together than in joints where watertightness is not necessary. In a lap joint, rivets must be so close together that the distance between a pair of rivets may not give a section of plate which is flexible enough to spring apart under the separating action of the caulking tool. Hence the reason for this necessity of very close rivet spacing on all parts of a ship that must be watertight.

The " grip " of a rivet is the distance between the head and point after it has been staved up. Thus, a rivet holding two plates together has a grip equal to the measure of the thickness of the two plates, plus a small amount as given later. The "pitch" of a rivet is the distance measured from center to center of rivets along the line of the riveting. That portion of the plate edges which lap and through which rivets are driven is spoken of as the "plate-landing." The end joints of the shell plating are overlapped. This type of joint is more efficiently caulked than the butt joint. The riveting of the shell landings on a ship binds the different strakes into one continuous surface.

Particular care must be observed to secure fair holes in three-ply riveting in the frame joints and gunwale bar.

If the holes are punched in the plates the rivets used in shipbuilding are slightly swelled in the neck, that is, that portion just under the head. This is necessary in order that the rivet may seat itself soldily in the punched hole when it is staved up. Holes punched in ship plating are slightly tapered, and a straight rivet is apt to leave an empty space within the riveted section. When a rivet is driven into a plate there is some danger of the head not being solidly set against the plate. To avoid this, therefore, the rivet should be well "laid up," that is, immediately after the hot rivet is placed in the rivet hole it is struck a few blows, either with a hammer or by operating the pneumatic holder-on to set the rivet well into the hole and the head against the face of the plate. The head of the rivet should not be struck after the final staving up, as this is apt to loosen it. As will be seen later, however, the point is finally finished after the rivet has been driven in place.

In working up the riveting, holes are sometimes found

which are not in line one with the other. Such holes are said to be "unfair," that is, they are out of line with each other. The riveter should never drive a rivet unless the holes are reasonably true. If they are unfair the bolters should be called and the plates drifted into proper position or "picked" out. The term "picked," as used in this sentence, applies to the chipping off of the edge of a hole with a gouge, so that the rivet may enter in the proper position. Holes which are out a great deal must be reamed larger. This work is attended to by the drillers and reamers. If all holes in the ship are reamed, the difficulty of unfair holes is largely done away with.

A half blind hole is one in which the hole in the plate on the inside laps over the outside hole approximately half of its diameter. When a hole is out of line to this extent it should always be drilled and reamed before riveting. Whatever the nature of the plating surface before riveting, it is rendered fixed and solid by the riveting process; therefore, it is important that the surface shall be properly aligned before rivets are driven.

All rivets used in modern shipbuilding are of steel. The various kinds of rivets are as follows: Pan head, snap or button head, flush or countersunk head, countersunk raised head, and tap rivets. Points are snap, hammered and countersunk. Pan and button head rivets $\frac{1}{2}$ in. in diameter or over have coned or swelled necks for punched plates, and straight necks for drilled plates. The advantage of swelled neck rivets is that the diameter of the punched hole on the die side of a plate is always slightly larger than on the punch side. In assembling, the plates are reversed, and thus with swelled neck rivets the holes are completely filled.

Rivets are made up of a "body," a "point" and a "head." The body is in the form of a bolt and the diameter of the body determines the size of the rivet. The head and point of the rivet vary according to the kind of plating which is to be connected by the rivet.

Fig. 5 is a Pan head rivet used for rivets below ½ in. diameter or for a larger rivet in a drilled hole and is used extensively in shipbuilding construction. This rivet possesses the important quality of strength and has a solid clamping effect on the plate. It is very easy to hold up during the staving operation.

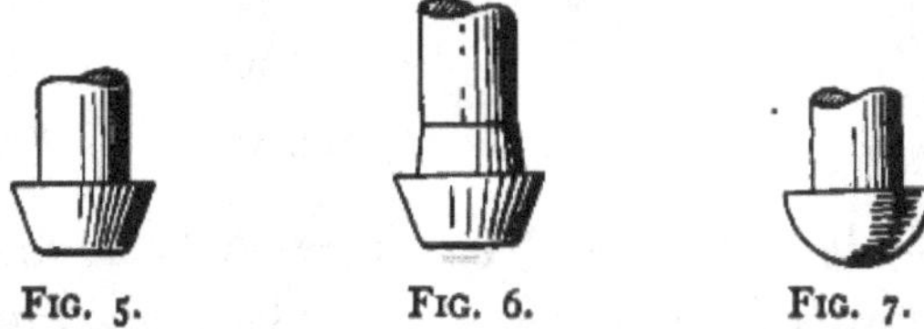

Fig. 5. Fig. 6. Fig. 7.

Fig. 6 is a "coned" Pan head rivet. This is used in all other places where a Pan head rivet is required.

Fig. 7 is a Button head rivet. This is used for casings and some foundations where flush work is not required.

Fig. 8 is a "coned" Button head rivet.

The countersunk head rivet (Fig. 9) is employed where flush surface is required, as on the gunwale. It is also used in many places where watertightness is very important because the making of a water-tight joint with a countersunk rivet is a comparatively simple operation.

The countersunk rivet, when finished, may or may not be perfectly flush or flat. When it is to be flush it must be chipped and finished off quite carefully. Where a perfectly flush surface is not necessary the rivet is allowed to set beyond the plate a small fraction of an inch as permitted by the specifications of the ship being built. The countersunk rivet is especially suitable for three-ply riveting, such as occur in many places on a ship.

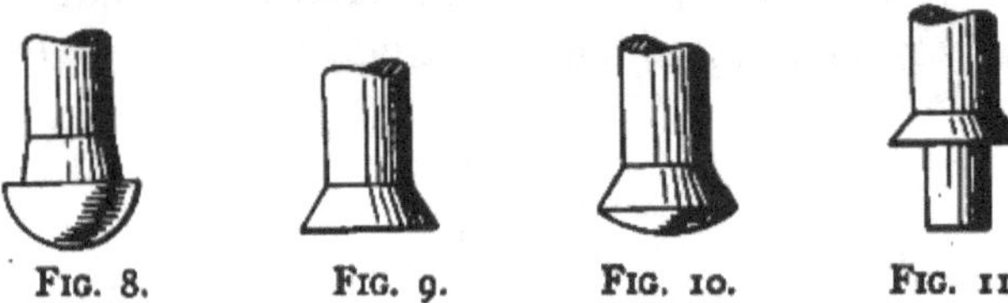

FIG. 8. FIG. 9. FIG. 10. FIG. 11.

Fig. 10 is a "raised" countersunk head and is used where the work is to be flush but the head must be caulked.

Fig. 11 is a "Tap" rivet. This is a threaded bolt with a countersunk head. Above the head a square end is left to turn the tap into the hole. After it is in place the end above the head is chipped off. These are used where plates or shapes are fastened to a casting.

Fig. 12 is a Countersunk point rivet used for flush work. When driving shell or deck rivets the finish is not exactly flush but the point of the rivet is drawn up a little in the center so it will project a trifle beyond the plate, to be sure there is ample material.

Fig. 13 is a Snap point rivet and is generally used for a Button head rivet.

The snap point is not extensively used in ship work, but since there are occasions when the workman must finish such a point, the following notes relative to this type of point will be of value. The rivet should first be staved up by holding the hammer so that it strikes directly on the end of the work until the shank grips the hole solidly. In this particular part of the operation the hammer is given a slight rolling motion. The rivet just finished is left, and a second rivet staved up as mentioned. Following the staving up of the second rivet the workman

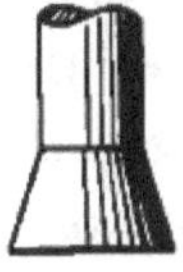

FIG. 12. FIG. 13. FIG. 14. FIG. 15.

returns to the first rivet and finishes by striking a few rapid blows after the work has cooled somewhat. This latter driving produces a nicely finished head.

Fig. 14 is a Hammered point rivet. This type is used as specially required.

Fig. 15 is a Liverpool or Oval point rivet. This is used on a finished, exposed surface with thin plating.

The proportions of the heads and countersinks vary with the different classification societies. There are no universal standards. Following are Lloyd's proportions: Countersink for plates in which countersink head and point is to be used should extend through the whole thickness of the plate when less than 14/20 or .7 in. thick; when .7 in. or above, the countersink to extend through nine-tenths the thickness of the plate.

The length for ordering pan and button head rivets

is measured exclusive of the head; for countersunk rivets and taps the ordered length includes the head to the top of the countersink.

Rivets are placed in a ship by a number of gangs. Each gang consists of a Heater, a Holder-on, and the Riveter. In remote places on the ship there is introduced into this gang an extra boy to pass the rivets from the Heater to the Holder-on, who inserts it in the hole ready for driving. Many times the Heater Boy throws the hot rivet through the air to the Passer Boy, who receives it in a "catching can" and then places the rivet in the hole. The duty of the Riveter is to drive or "stave" the rivet point up solidly against the plate so that the two parts of the plate are firmly jointed together. In shipbuilding practice this is done by means of an air hammer, which is the implement used by the Riveter in doing practically all of his work. It is operated by compressed air, fed to it by means of the hose which is attached. The riveting tool, or "die," as it is known, is placed in the socket of the hammer and held firmly against the point of the rivet, which is rapidly driven into place. The hammer is started and stopped by pressing the air valve lever, directly under the operator's hand, in the handle of the hammer. The driving end of the hammer is supported in the left hand of the operator, while he uses his right hand for controlling the air valve and pressing the hammer firmly against the rivet to be staved.

One of the particular points which must be gained by experience is the amount of driving necessary to firmly join two plates.

It is the duty of the Heater and Holder-on to send through the right kind of rivets as regards length, and it is also the duty of the Riveter to be sure that the rest of his gang are doing their share. In this way it is taken for granted that the work has been thoroughly prepared and well bolted, but every Riveter should be on the lookout to notice whether or not there are sufficient bolts in the work and if the plates have been "hove-up" tightly one against the other, otherwise the rivets are liable to squeeze out between the plates and the work will have to be done over again.

It is well for the Riveter to also be on the lookout to see that all tapered and parallel liners are in place before he drives rivets through the holes where these should be fitted.

Work on the tank top or deck is usually thought to be the easiest, with the bulkheads coming next, followed by the side shell, and the bottom shell being the hardest of all. When working on the decks it is customary for the Riveter to drop down on his left knee (if he is right-handed) and keep the inside of his ankle and his left hand as a guide for the lower end of the hammer. Where the work on deck is located against a bulkhead or the ship's side, it is better to move the foot so that the hammer comes against the outside side of the shoe, thus putting all of the body off to one side of the rivet hole rather than directly over it, which allows ample room for working the hammer and yet having it as close to the vertical plate as needed for the work. In this position it is well to keep the elbow in close to the body so that there is not too much leverage of the arm, the kick of the hammer being carried up through the body and not felt to so great an extent.

When driving on the bulkhead or side plating, the work varies according to the heights of the rivet holes, and the Riveter is obliged to work from the height where he is on his knees until the rivet holes are about on a level with his shoulder. When these are driven he should build staging in order to reach for those above. This same method of carrying the arm close to the body, having the body slightly leaning toward the rivet hole is the best position for this kind of work, although when the riveting is about knee high it is best to brace the hand against the inside of the knee and take all the jar on the leg. When driving countersink rivets on the shell the necessary rolling motion of the gun can be given by swinging the leg.

When driving button set rivets the end of the gun need not be so near the hand as is necessary when flush rivets are being driven. For work on the bottom shell, or "upper-cuts" as it is often termed, the riveter generally needs a block or box on which to rest his foot. If he is right-handed he will place his right foot on this block to rest his elbow sufficiently so that it is possible for him to get a good brace with his arm, the back of his left hand just touching the plating and the end of the gun being held loosely between the fingers and thumb. This gives ample backing for the kick of the hammer and a guide for the hammer around the rivet point, as most of this work is countersunk point.

The Riveter naturally falls into three different types of driving: for the button set the drive is straight, forcing the material into the hole and forming a round point; for the hammered point, the workman will fill the hole and then form a little tent-like point on the rivet, spreading it out over the plate adjacent to the hole; for the countersunk point the riveter will be sure that the hole is filled and will then work the point into the countersunk circle around the edge of the opening and force the material up toward the center, so as to have the point extend a full one-eighth of an inch above the plate.

Care should be taken when driving this type of rivet to avoid striking the plate around the hole, as all the force of the hammer should be expended on the material of the rivet.

On bottom shell and tank tops it is usually customary to drive long rivets and then chip off the excess "rag." This is done so that there may be ample material for caulking afterwards, in case the rivet is not tight or should leak. For riveting on the side shell or on the decks this is not necessary providing care is taken to have the rivet the proper length for a good point. When driving chipped rivets it is well to drive one, chip it, then go ahead and drive the next, chip it, and then go back on to the previous one and finish up the point, repeating operation in each case. This is done to be sure that the rivet is well hammered up against the plate, as many times, due to shrinking of the metal as it cools, it is liable to loosen a little.

After the rivet has been thus hammered up there is little danger of a leaky rivet. On deck work and other places where the workmanship is not so particular as on tank tops and bottom shell, it is the custom in many yards for the Riveter to drive the rivet and finish it complete before he goes on to the next one, taking it for granted that the opening has been filled and that the rivet will be tight.

Heating the rivet properly is important. If the rivet has not been heated sufficiently, it cannot be driven solidly into place, while if it is too hot it is burnt, and therefore useless. When the Heater is working at the furnace, he keeps a large number of rivets in the heat at one time. He should place the points of the rivets down so that they are heated quite hot, while the heads are not heated as hot. This is necessary in order that the point may be soft, while the head will be hard in order to hold in place during the staving process.

The rivets are handled by tongs which enable the workman to pick them quickly from the furnace, pass them to the required position and insert them in the holes punched in the plate. After the rivet has been properly heated, it is passed by the Heater to the Holder-on, who places it in the hole and immediately places against it the "holder-on." A tool known as a "dolly" is used on some work instead of the holder-on; the dolly is a hand tool, which serves the same purpose as the pneumatic holder-on.

Rivets are staved up as follows: the hot rivet is inserted in the hole and the pneumatic holder-on or the dolly placed against the head. After the rivet is placed in the hole the holder-on is operated in order to lay the rivet up and give the head a good bed on the plate. It will be seen that the holder-on prevents the rivet from being driven out of the hole during the process of staving. At first the rivet is struck by the hammer fairly on the end, so that the shank may be staved up, and fit the hole throughout its length. If it is a countersunk point and the rivet shank is rather long, which is usually the case, it is then struck to one side, slightly, so as to bend it over in such a way that the extra material not required to fit the countersink may bulge to one side and thus be readily chipped off while it is hot. This piece of the rivet which is pushed to one side is known as the "rag." After thus staving up the rivet the Riveter lays the riveting hammer aside and takes his chipping hammer and chips off the rag. He then takes his riveting hammer and gives it a second series of blows, delivered with care, around the edge, so placing his hammer that the rivet raises slightly towards the point. This process thoroughly caulks the rivet and finishes it off smoothly. If the Riveter is skillful it will be difficult to tell exactly where the rivet is located in the plate because of the very smooth finish thus produced.

After the first series of blows has been made, the rivet point shows a slightly bluish color. If the second series of blows is given at the proper time, this bluish color will entirely disappear and leave the rivet the same color as the adjoining plate.

In working during hot weather when the rivet cools slowly, it is wise to delay giving the final series of blows until another rivet has been clinched.

In order to insure a perfectly tight joint it is a good plan, when giving the final series of blows, if the plate also be given a few blows along the edge, between the rivets, so as to bring up the faying surface everywhere into close contact. In placing shell riveting, it is customary to assign a section or "berth," as it is termed, to each riveting squad. The usual procedure is to rivet each landing, one after the other, beginning with the garboard strake, being sure that this strake is properly faired up before starting riveting. The rivets are then hammered up one after the other, the frame rivets and butt joint rivets being taken in their order. The Riveter should be careful that he rivets up only those frame rivets located from center to center of each strake. That is, he should not rivet for a long distance up a frame, leaving a number of butt joints, to be riveted between the frames.

In order to secure good riveting on water-tight work, many ship inspectors will insist that the Riveter use rivets which are too long and then cut off the excess stock. This is done to be sure that there is plenty of material to fill the hole and that the body of the rivet will be swelled out enough to do it and still have enough left to form a good, raised point.

The general rule for a "raised" point is to have the point of the rivet just high enough above the surface of the plate so that it is possible to "rock" a rule across it, allowing the ends of the rule to touch the plate when the rule has been tilted. A two-foot rule folded to a six-inch length is the common trial length.

This system of having extra stock on the length of the rivet is not followed so commonly on the "West Coast" of the United States as it is in "the East." The riveters there try to have the rivets of the correct length and then drive and finish the rivet with one hammer and at one operation.

It is customary for these men to have about six rivet dies at all times. When the workman starts in the morning these dies are all ground to a concave surface of about one-eighth inch depth, with a sharp edge. These dies are longer than those used for driving, when working with a chipping hammer, to remove the "rag," to allow for them to be worn down during the grinding.

To give the necessary force to these heavy dies a No. 80 riveting hammer is used. As these hammers are heavy to handle the workman wears a harness consisting of a leather strap over each shoulder down under the arms; these meet in a pad at the back something like a pair of suspenders. From this pad two straps are carried around under the left arm (for a right-hand man) and in front of the body they terminate in a ring. From these two rings a rope is carried, starting at one ring, passing to near the right side and then back to the other ring. A pulley is run on this rope and at the end of the pulley a spring and beyond it a hook is carried. This hook is used to catch onto the plug of the hammer where the air hose is attached. In this way much of the fag and strain, due to the rebound of the hammer, is taken up by the spring; the pulley working along the rope allows freedom of action and the attention of the workman is given to holding the weight of the hammer and swinging it around in the necessary circle to finish the rivet point.

As the hot metal is driven into and is filling the holes the workman will force it down evenly and then by carefully going around the edge of the rivet will drive it down without marring the surface of the plate itself. This work of finishing requires a skillful Riveter and it is because of this that many of the inspectors are not in favor of allowing this style of riveting to be done in their yards, as there is a tendency to jump off the rivet onto the steel plate and cut it up with the sharp edge of the die. The system has its good and bad points and is only mentioned in passing, not argued for nor against.

For first-class work all plates should be punched for rivet holes from the "faying surface" (which is to be placed against the surface of the other plate). This should be done because the punch makes a hole in the plate which is not parallel-sided but is partly cone-shaped, due to the crowding action of the die.

When the plate has been punched in this manner, a "cone" neck or "swell" neck rivet should be used to fill the opening to good advantage.

Many times this would necessitate the turning of the plate while in the Plate Shop, a proceeding the workman is apt to avoid, with the result that often the burr which occurs around the edge of the hole as the punch comes out through the plate is on the faying surface (a condition to be avoided).

Because of this, cone-neck rivets are not used so largely as they otherwise would be, as in such a case the cone part of the rivet would be on the wrong side of the opening. Under such conditions it is best to use a straight rivet and fill the cone portion of the hole while forming the point of the rivet.

When a number of plates, exactly alike, are being fabricated they are clamped together, a mold put on the top one and the rivet holes are then drilled through all plates at once, making the holes all straight-sided and cone neck rivets cannot be used.

Fig. 16 shows a "punching" just falling from the plate. The effect of the die is to bend the face of the punching and widen the diameter as it passes through the plate.

Fig. 16.

Fig. 17 shows a straight rivet in a punched hole. Note the space on each side of the neck.

Fig. 18 shows a cone-neck rivet in the hole, the shape of the rivet being designed to quite fill the opening in the plate.

A number of different sizes of hammers are available for different classes of work, as will be seen in going over the table on page 76. There are a few important

Fig. 17. Fig. 18.

features relative to the care of the hammer to which the operator should pay attention. Do not roughly use this tool, as some parts of it are very accurately made and the serviceability is greatly increased by its careful handling. Pneumatic hammers must be properly lubricated in order to give desirable results. One of the most important factors connected with their care is to keep them clean and well lubricated; when working constantly one should disconnect the air hose once in three hours and squirt some oil into the air hose connection. If the hammer does not work properly, report it to the foreman in charge of the gang, and let him give directions as to what should be done in order to keep the tool in good condition. As the air taken into hammer contains some particles of grit or dust, it is impossible to prevent this matter from entering into the working parts and thus doing damage.

The use of a poor grade or heavy oil will cause trouble in the operation of the hammer. A good plan to follow in cases where oil which is too heavy has been used for lubrication, is to clean the hammer thoroughly by the use of benzine or gasoline, which is put into the hammer through the hole in the end where the air tube connects. (The air tube connects on the lower part of the handle of the hammer.) This loosens all foreign matter and cuts the thick oil, which then can be removed by blowing the air through the hammer. It is an excellent plan to put the hammer in a bath of benzine over night occasionally, then blow out any dirt by connecting the air tube in place, thus cleaning the hammer thoroughly. Such treatment not only prolongs the life of the tool, but makes it much more efficient in use. The workman should keep in mind that the riveting hammer is made with all the parts proportioned to meet the various requirements of the work to which these tools are adapted. Occasionally a workman will attempt to make some changes in the parts of the tool, a practice that is decidedly objectionable. The parts of the tool should be used as furnished by the makers, and when parts are worn out they should be replaced with new ones supplied from the manufacturers of the tool. Follow this plan and long and efficient life is assured this class of equipment.

The most flagrant violation of the proper use of pneumatic hammers on the part of workmen is to introduce short pistons in the working chamber. These are rapidly worn out and cause a great deal of trouble, not only for the yard, but for the workmen as well. The reason why workmen sometimes follow this practice is because, for a brief period, the hammer will do a greater amount of work. This period of extra service is very short, and the tool is often ruined, due to the improper use mentioned.

There are available automatic oilers which may be used to advantage on these hammers. These may be used on all classes of this equipment. The table on page 76 will give a good idea of the service, which may be expected from various sizes of pneumatic hammers.

This table will also be of assistance in deciding the class of work which various kinds of hammers will do.

In the use of pneumatic hammers, a tool is placed in the end for doing the actual driving of the rivet. These tools are known as sets, and they are of various sizes to accommodate the different rivets.

In the stem and stern frames of the ship, very long rivets are used. Where this is the case, the rivet must fit the holes very closely and be driven in place. Such long rivets are heated only at the point and never in the body. The Holder-on after the rivet has been brought

Style	Size	Diameter of Piston	Length of Stroke in Inches	Cu. Ft. of Free Air Per Minute	Work to which this Tool is Adapted	Length Tool in Inches	Size of Hose Connection
Long stroke riveting hammers	#90	$1\frac{1}{16}$	9	25	Driving rivets up to $1\frac{1}{4}''$ diam.	$23\frac{1}{2}''$	$\frac{3}{8}''$
" " " "	80	$1\frac{1}{16}$	8	25	" " " " $1\frac{1}{8}''$ "	$22\frac{1}{2}$	$\frac{3}{8}$
" " " "	60	$1\frac{1}{16}$	6	25	" " " " $\frac{7}{8}''$ "	$20\frac{1}{2}$	$\frac{3}{8}$
" " " "	00	$1\frac{1}{16}$	9	25	" " " " $1\frac{1}{8}''$ "	$23\frac{1}{2}$	$\frac{3}{8}$
" " " "	50	$1\frac{1}{16}$	5	25	" " " " $\frac{3}{4}''$ "	18	$\frac{3}{8}$
" " " "	40	$1\frac{1}{16}$	4	20	" " " " $\frac{9}{16}''$ "	17	$\frac{1}{4}$
Chipping and caulking hammers	#0	$1\frac{1}{16}$	5	20	Ex. h'vy chipping and caulking	$18\frac{1}{2}$	$\frac{3}{8}$
" " " "	1	$1\frac{1}{16}$	4	20	Heavy chipping and caulking..	15	$\frac{1}{4}$
" " " "	2	$1\frac{1}{16}$	3	20	Medium chipping and caulking	13	$\frac{1}{4}$
" " " "	3	$1\frac{1}{16}$	$1\frac{3}{4}$	15	Light chipping and caulking ...	12	$\frac{1}{4}$
" " " "	A	$1\frac{3}{16}$	3	20	Heavy chipping and caulking ..	15	$\frac{1}{4}$
" " " "	B	$1\frac{3}{16}$	2	15	Medium chipping and caulking	$12\frac{1}{2}$	$\frac{1}{4}$
" " " "	BB	$1\frac{1}{16}$	$1\frac{1}{2}$	12	Light chipping and caulking ...	$11\frac{1}{2}$	$\frac{1}{4}$

to the required heat, sometimes cools the body of the rivet in water before driving it into the hole.

Rivet Testing

All rivet work must be tested in ship construction in order to be sure that the rivets are tight and sound. The closeness of a joint may be tested by the means of a thin-bladed knife, the insertion of which between the faying surfaces should not be possible. In poor work the testing knife may be entered freely into the joint. This should not be possible at any portion of a riveted joint. A tight rivet can be tested by striking with a light hammer. If the rivet is not tight a jarring noise can be heard, and the hammer will not give a sharp rebound. A tight rivet when struck with a hammer always gives a sharp, clear, ringing sound.

In testing a rivet the workman places his finger against the head and plate, and hits the rivet a few light taps. The sound, together with the sense of touch, after a very limited experience, will enable the worker to tell at once whether or nor the rivet is tight.

A defective rivet may sometimes be made good by heading up slightly while it is cool. If a few blows do not produce a solid rivet, however, the rivet should be "backed" out and a new one put in its place.

Tap, stud, or screw rivets are used in riveting the shell plating to the stern frame. This is not a desirable form of rivet for a general run of work. Tap rivets are screwed in place by means of a wrench in a manner similar to that in which a bolt is tightened up. They should fit the hole in which they are placed tightly, so that there may be no chance for them to loosen in their seat. The holes for tap rivets are usually countersunk in place on the ship and the rivets are screwed tightly into place, after which they are staved up in a manner similar to that for the ordinary clinched rivet, in staving use the hammer so as to tighten the rivet, striking to turn rivet deeper into the hole.

For the purpose of screwing up a tap rivet a square protection is left on the head. Tap rivets are used in places where it is not possible to work from both sides of the joint in riveting. A great many of the rivets used about the stern post of the ship must be of this type.

When staving up a rivet the Riveter chips off the extra material left over after the staving up process. The rivet should never be hammered down and spread out over the outside surfaces of the plate.

ALLOWANCE FOR POINTS IN LENGTH OF RIVETS WITH TWO THICKNESSES CONNECTED

	Diameter of Rivets					
Type of Point	1/2	5/8	3/4	7/8	1	1 1/8
Countersunk	1/2	5/8	3/4	7/8	1	1 1/8
Hammered	1/2	1/2	1/2	5/8	5/8	5/8
Snap	7/8	1	1 1/8	1 1/4	1 3/8	1 1/2
Oval	7/8	7/8	...	...	...	...

The above allowances are based upon the average practice at various United States Navy and private shipyards.

Rivets are shipped in kegs of 100 and 200 lb. In

ordering, the diameter should be given first and the length following, thus: $\frac{1}{2}$ in. by 3 in.

In riveting the workman must be sure that he selects a rivet long enough to form the point he wants to make. If the rivet is not long enough the point cannot be formed. If it is too long, stock is wasted. The table just given will enable one to quickly determine the length of rivet to form a point. The plain rivet is the snap form. For special rivets, instructions should come from the design department, giving amount of stock to be allowed for heads.

Length of "grip" is found by adding thicknesses of pieces to be riveted plus $\frac{1}{32}$ in. for each joint between plates.

Snap rivets should have points true to hole in which they are driven. Variation from truth can be detected by measuring the pitch.

The Inspectors should not permit much redriving of cold rivets to tighten them nor allow excessive caulking of rivets.

Pits, rough surfaces and cracks in a rivet indicate overheating, though these are not absolute proof of overheat. The inspector, when finding this condition should order a few of the bad looking rivets cut out; if the rivets do not yield readily to the "buster," if the break shown by cutting of the rivet head is clean and shows a good surface, the evidence of burning is misleading.

The detection of loose rivets has been described. Loose rivets are made to appear tight by going around the edges with a caulking tool. The inspector should look for caulking tool marks, and be guided by instructions from the office as to the amount of caulking he is to pass. Loose rivets on snap work are made to appear tight by side driving with the snap. Such driving leaves a ridge around the head, and when such ridges are found, the workman should be watched. These may not be loose in all cases, but observation of the workman will show whether it is tightening of loose rivets, or finishing the head that causes the suspicious marks. A few rivets may be cut out if necessary, in order that the inspector may determine the quality of the work.

Rivets should be watched on the "held up" side. A rivet may show perfectly tight on the point, but be loose on the head. Any marks on the head, or held up side of a rivet, indicate an effort to remedy some defect.

Countersunk rivets often give trouble, because the blank may be slightly long, and the excess material is allowed to spread out under the set, and overlaps the hole. This edge if not quickly chipped off, will give the appearance of a tight point, when really it is loose.

In marking rivets to be cut out, use a center punch, making a deep mark. Paint the rivet, and put a mark on the plate beside the rivet, so it can be quickly located for a second inspection.

In selecting dies to do different classes of work the Riveter should remember that the same die is not used for all sizes of rivets, he should ask for a die that will fit the rivet to be driven, both for the holder-on and for the hammer. Dies are of various shapes, those for snap points being different from those used for hammered points.

Notes on Riveting

This is one of the most important parts of shipbuilding. At the present time a small vessel has been launched in Great Britain for which it is claimed that all the seams and butts are electric welded instead of riveted. It is not safe to condemn any new thing, but at the time this is written riveting still is the only method in general use for connecting ship plates.

Details of the ship construction stand or fall according to the design in the drafting room and the care with which the work is done on the ship.

Two of the principal things are rivet spacing and type of rivet used.

Spacing of rivets is given in terms of the diameter of the rivet, as it bears a ratio or percentage of the whole joint.

The closest spacing is three diameters and the largest is eight diameters. All the different conditions in the ship are met between those two limits.

Oil-tight work requires the rivets nearest together as oil will seep through wherever possible and must be well caulked to prevent it, hence close spacing of rivets to hold the plates together.

Water-tight work calls for the next step up in spacing, being about 4½ in. diameters. (This means 4½ in. for a 1-in. rivet.)

Tank top, etc., come next with about 5 diameters.

Foundations require about 6 diameters.

Floor brackets, intercostals, etc., require about 7 diameters.

Beams to deck plating and non-water tight frames to shell plating take 8 diameters.

Plates are often punched the full size of the rivet. After the plates are together and regulated they are reamed out to full size of the rivet. Sometimes the holes are punched small and then reamed to suit rivet.

Rivets below 3/8 in. diameter are driven without heating but all above that size must first be heated before they are driven.

In Lloyds Bureau of Shipping the standard thickness of a 1-in. plate is 20/20. Therefore a 1/2-in. plate is 10/20, 1/4-in. plate is 5/20.

The United States naval practice takes a plate 1 in. thick and 1 ft. square as being 40 lbs. in weight. A 1/2-in. plate is 20 lbs., a 1/4-in. plate is 10 lbs., etc.

As much of the work done in the United States is according to the American practice it is well to become accustomed to the method of measuring by weight.

The following table may be taken as a useful standard:

1/4	in.	rivet	requires	up to	2 1/2	lb.	plate	also	an	angle	with	1	in.	flange.
3/8	"	"	"	3	"	5 1/2	"	"	"	"	"	1 1/2	"	"
1/2	"	"	"	6	"	7 1/2	"	"	"	"	"	2	"	"
5/8	"	"	"	8	"	12 1/2	"	"	"	"	"	2 1/2	"	"
3/4	"	"	"	13	"	19 1/2	"	"	"	"	"	3	"	"
7/8	"	"	"	20	"	29 1/2	"	"	"	"	"	3 1/2	"	"
1	"	"	"	30	"	39 1/2	"	"	"	"	"	4	"	"
1 1/8	"	"	"	40	"	50 1/2	"	"	"	"	"	4 1/2	"	"

This table should be fixed in the memory as it forms a basis for estimating required sizes. Having given the thickness of plate one can know at once just what size rivet is required and what size of an angle. (As a general thing the thickness of the angle is about the thickness of the plate in order to be of service.)

When two plates are joined the size of rivet called for varies if they are of different thickness, according to the work to be performed. If it is for a strength joint, then use a rivet for the thicker plate; if it is for an oil- or water-tight joint then use the rivet as required for the thinner plate.

Riveting is generally more than single riveted when connecting two plates.

Fig. 19 shows double, chain riveting. If there were a third row it would be treble, chain riveting, etc.

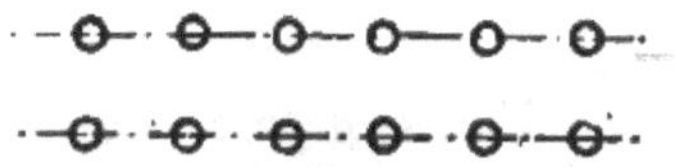

FIG. 19.

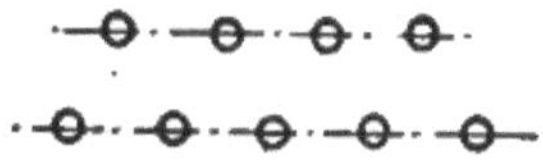

FIG. 20.

Fig. 20 shows "staggered" or "zigzag" riveting. This is often used when an angle with a broad flange is connected to a plate.

Fig. 21 is an angle with unequal legs. When one leg or flange of an angle is against a plate the other is called the "standing" flange. When an angle of unequal legs is shown on a blue print, the flange which is shown laying flat on the plan is read first, i.e., 3½ in. by 4 in., or 4 in. by 3½ in. according to which is facing you.

All angles are formed at 90 degrees and when necessary for ship work they are heated and then "beveled."

Fig. 22 shows an angle at 90 degrees, dot and dash

line is an " outside " bevel, the dotted line is an " inside " bevel. Inside bevels are to be avoided as they are very difficult to rivet.

The outside point of the angle is called the " heel " and all measurements are taken from the heel. The two ends are called the " toe " of the angle. The inside, opposite the heel, is called the " bosom " of the angle.

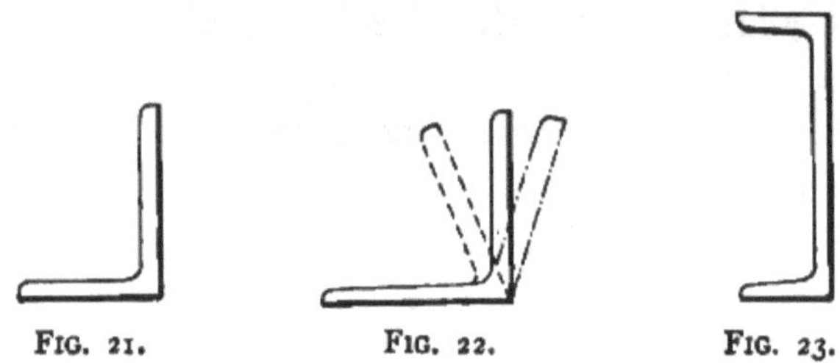

FIG. 21. FIG. 22. FIG. 23.

Fig. 23 is a channel bar. The long part is called the " web " and the two smaller parts are called the " flange." The channel also has a heel, toe, and bosom.

Angles and channel bars are the two most common shapes now used in merchant shipbuilding. There are other " shapes " (a distinguishing definition from " plates ") used in naval work but they will not be mentioned here.

DUTIES OF THE HOLDER-ON (SOMETIMES CALLED " BUCKER-UP ")

In ship riveting gangs the Holder-on very often acts as the mainspring for the whole gang, as it is due to his foresight in planning his work that the speed of the gang shows to good or bad advantage,

The duty of the riveter is to drive the rivet point after it has been shoved through the plating and is being held in place. He must simply form the rivet point so that it is well done. The Holder-on must receive the rivet, force it into place, and hold on tightly while the point is being formed. This involves a number of different operations and the use of varying tools.

The tool used for holding on varies according to the location on the ship where the gang is riveting and the kind of rivet being driven. One tool often used for this work is the "dolly bar," which is a solid mass of cast iron, having a block-like end with a short handle terminating in a ball for easy holding with the hand. This weighs about thirty pounds and is simply held in position, the force of the blow on the rivet driving it back a little but its weight helps to overcome the force of the blow. These dolly bars have either a straight handle or a bent handle (called a "goose-neck"), the face on the end of the block being either square across or formed at a bevel so that the bars can be used under a number of different conditions. These bars are generally used in confined spaces where it is difficult for the Holder-on to get in close to the rivet with other tools.

Air hammers are generally used on bulkhead and shell work for holding on; the hammer working against the riveting hammer on the other side, giving a heavy blow to the rivet.

For deck work where the Holder-on is directly beneath the deck and in other similar conditions, it is customary to use the "air jam." This is a pneumatic tool mounted on a pipe extension, the bottom of which rests on staging or some other support. In both the air hammer and air jam the die is used in the same way as they are fitted in the hammer for riveting. These dies are usually formed for buttonhead or panhead rivets made to fit the shape of the rivet head, for holding the rivet in place and not deforming the shape of the head.

The Holder-on should be careful to secure the proper die for the kind of rivets that are being driven and avoid using a die not adapted to the head on which he is working. The Holder-on operates the air jam by putting it directly under the rivet after the rivet has been placed in the hole and turns on the air through the handle valve, and this air pressure shoots the die up against the rivet head and holds it firmly in place. The air jam should be held very close to the rivet, and this is done by fitting a long pipe extension, the length varying according to the location of the staging or backing which is being used.

The "combination" or "vibrating air jam" is very popular with many holders-on. This tool is similar to other air jams except that it has a plunger which operates in a manner similar to a riveting hammer. When the rivet is in place the air jam is set in motion and the hammer works on the rivet head, against the rivet point, in the same way that work is done on bulkheads, when the Holder-on is using a riveting hammer fitted with dies to suit the head of the rivet.

Air jams are commonly used wherever it is possible to secure a firm footing for the end of the pipe extension to take the strain of holding-on off the workman, the thrust from the blow of the riveting hammer passing down through the machine to the support. The Holder-on merely sets the machine in position and starts it to work. Air jams are released by means of a valve which relieves the pressure and the plunger drops back into the machine.

Work on the tank top is considered the most difficult for the Holder-on as he must travel through the Inner Bottom compartments, holding-on against the under side of the Tank Top plating. On other parts of the ship he usually has the same advantage of light and air as the Riveter.

One of the tightest places for the Holder-on to work is when riveting down through the margin bar of the Tank Top at the ends of the ship, where a close bevel between the tank top plating and the side shell gives a very small space in which to handle the tools. This is often overcome if the foreman of the ship is wise enough to first regulate his plating and bar and then draw the margin strake of plating a few feet inboard toward the center line of the ship, thus allowing the Holder-on more room and putting the work in a position where he can have it directly over-head.

It is the duty of the Holder-on to see that his staging is prepared in advance so as to avoid holding up the entire gang while the ship carpenters are putting in other planking from which he can work.

The size and proper length of the rivets should be a care of the Holder-on, as he is better able to judge the length required because he can see the framing of the ship which is hidden from the Riveter. He should also have a plentiful supply of rivets ahead of the job rather than depend upon the Heater Boy who is not always awake to his responsibilities.

Sometimes it is necessary for the Holder-on to refuse rivets passed to him because of the burnt ends. These should never be used as the rivet which is a little burned is in a crumbled condition, and the Riveter will not be able to form a point from material under such conditions.

One fault often found with Holders-on is that they are not careful enough to have the under side of the rivet head hard up against the plate. This is necessary for good workmanship and especially for water-tight work.

Duties of the Heater and Passer Boys

Rivets are heated in small movable hearths. Either coke, gas, or oil is used for making this fire, and a blast is obtained from the compressed air tank which produces a hot flame. In heating a rivet, some Heater Boys use a punched plate, laying it over the top of the hearth, into which the rivets are laid, during the heating process. Such a plan assures efficient heating and is very convenient, especially for beginners. After considerable practice such a plate is hardly necessary, however.

The coke fire is most commonly used and is also the most difficult to handle. "Foundry" coke is best for use in rivet fires, gas coke not being suitable, and it is best to break the coke into small pieces if it is not supplied that way. In starting the fire, a plentiful supply of small pieces of wood in the fire pot is helpful. The coke should be kept out well around the edges of the fire so that it will burn evenly and only a moderate draft should be used, as forcing the draft doesn't make a hotter fire. The length of hose should always be sufficient to permit moving the forge without delaying the work of the gang while additional hose is being obtained and attached.

It is necessary often to clean the fire and this is done by removing with the tongs the large lumps of clinker and then turning on the air for a very heavy draft, thus forcing the small clinkers and ashes up into the air and clear of the fire. Additional coke is then added to the hot pieces left on the bottom of the forge. When throwing out the large clinkers, they should be gotten rid of and not mixed with the supply of new rivets, as much time will be saved by keeping the rivets clean.

It is easy to see that upon the enterprise and initiative of the Heater Boy, the speed of a riveting gang depends, and he is usually kept moving by the "Holder-on." A good Heater Boy will recognize at a glance the diameter of a rivet (the comon ones are $\frac{5}{8}$ in., $\frac{3}{4}$ in., $\frac{7}{8}$ in. and 1 in.) and he will know that their length is measured from under the head. Countersunk rivets are measured from the extreme end. He will try to furnish the correct rivet as called for.

When first learning to heat rivets generally about five are placed around in a ring, with the heads just peeping through the coke. As soon as the first rivet is hot enough to use, it is drawn out with the tongs, and the hole left in the coke is filled up with another rivet. The second is next drawn and the hole filled up before taking the next one, and so on. With a little practice the Heater will be able to cover up the heads more and still be able to find the rivets when wanted for use. The tongs should be kept out of the fire as much as possible so they will not become hot, which is bad for the tongs and bad for the hands. When called away from the forge for any reason the air should be turned off so that the forced fire will not burn the rivets.

The learner should try to practice as much as possible in the spare time, the style of throwing the hot rivets. With feet well spread apart the tongs holding the rivet swung back and under the body as far as can easily be done, then the body is thrown back and straightened up at the same time swinging the tongs up and toward the man who is to catch the rivet, letting go with the tongs at just the right time so the rivet will travel in line and toward the Holder-on or Passer Boy, who has his catching can ready.

The Passer Boy should practice catching the hot rivets in his catching can and should learn to stand so that the can is beside him and then his body is not in line with the moving rivet. If the rivet misses the can it will pass harmlessly (unless it hits someone else further on) but if the Passer Boy holds the can in front of him and makes a poor catch, his body is liable to stop the rivet, with painful consequences. Keep out of line of the Holder-on, too.

When taking the rivet out of the catching can it should grasped by the head so that it is possible to put it into the hole at once without making another stop to regrasp it, as would have to be done if taken by the body the first time. The short "passing tongs" are used. A right-handed boy holds the can in his left hand and the tongs in the right hand.

The Passer Boy should help the Heater Boy with his rivet supply and starting the fire.

Duties of the Caulkers' and Chippers'

Caulking

The Chipper trues up all irregular plates, cuts off plate edges that are too long, chips bosom pieces and does similar work.

The Caulker tightens up all work that must be water-tight. This work is done on shell plate edges and butts, water-tight floors and bulkheads, and around the decks.

For chipping and caulking a hammer of the same design as the riveting hammer is used. The tools have a hexagonal shank which fits in the hammer. Two kinds of chisels are used, the " cape " and " broad " chisel. The broad chisel is used on open work, the cape for getting in corners and close places. For most ship work the chisels are " side ground."

The principle of operation of the hammer is described in Chapter III. The gun should be held firmly in the palm of the hand, so that the workman can operate the trigger with his thumb, lightly or with more force as is necessary, but at all times maintaining a firm pressure on the handle of the gun in order to hold the loose tool in the other end firmly against the material which is being worked.

In caulking, the lower edge of a plate is forced against the surface of the one next to it so tightly that the joint actually becomes metal to metal and forms a barrier to any leaks. Oil will sometimes find its way through a crevice which ordinarily will prevent water from passing, so that particular care must be given to all oil-tight work. In "staving up" the edge of one plate it forces it into contact with the surface of another. Caulking may injure a joint if improperly done; therefore, a caulker should be careful not to spread out the edge of the plate. In caulking, a "splitting tool" is first used so as to make a narrow furrow along the edge of the plate about $\frac{1}{16}$ in. deep. Then a setting tool is used which completes the caulk by squaring out the groove. The setting tool is practically the splitting tool reversed. This square groove forming the "caulk" is from $\frac{1}{8}$ to $\frac{3}{16}$ in. wide and $\frac{1}{16}$ in. deep. The first width is used on thin plates, and the latter on the heavy plating. If very nice work is wanted, a third tool known as a "finisher" is used. It is a regular caulking tool, and is applied lightly along the caulk. In caulking, the shoulder should be deep and square. To produce this deep shoulder a heavy caulking hammer must be used.

The splitting tool has a beveled point so that it is possible to work the material down from the beveled edge in toward the joint between the two plates (Fig. 24 *A*) and then by turning the tool over it is possible to go right over this joint and force the material close into the crack, thus closing up the space. (Fig. 24 *B*.) It is necessary to hold the caulking tool at such an angle that the metal is forced back against the edge—leaving the caulked line smooth and straight.

Sometimes the crack between the two plates to be caulked is wider than it might be in other places, although the plates may be tightly riveted together. It is necessary to look carefully before starting to caulk to see what condition must be met, so as to determine whether a broad or

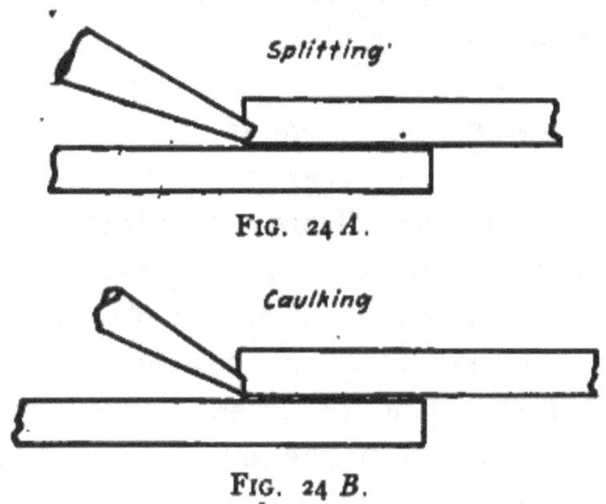

FIG. 24 *A*.

FIG. 24 *B*.

narrow caulking tool should be used. It must be understood that caulking is a means of preventing water from seeping through between the plate and angle or between two plates.

When caulking tapered liners, the butt end of the

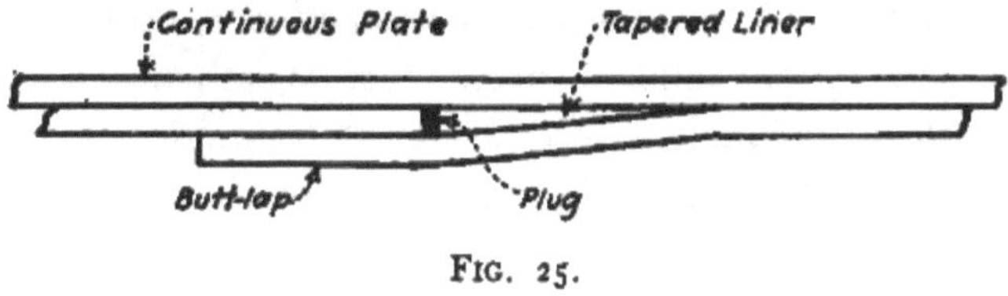

FIG. 25.

liner often does not meet the end of the adjacent plate, leaving a large opening which it is necessary to fill by means of a small metal plug (Fig. 25). If the opening is small enough it is possible to use the roughing tool, the beveled end of which is used to break in the upper corners of the adjacent edge of the plate until they meet, thus forming a wedge and closing the opening. The roughing tool is then used to hammer these together, forming a tight joint.

Three conditions are met with in caulking liners. First, where the liner has its end in line with the edge of the adjacent plate. Second, where the liner is not out as near the edge as it should be. Third, where it stands out too far beyond the edge. The first case is easily handled by caulking the adjacent end of the liner and the plate tightly against each other. In the second case it is customary to chip the plate back a little beyond the extent of the liner in order to have the liner and plate on line to make a good job with the caulking tool. In the third case it is customary to caulk the liner against the plate before chipping off excess material of the liner then the edge is caulked in the usual manner.

When caulking an angle clip against a bulkhead or other water-tight plate the toe of the bar which is against the plate is chipped back for a short distance to take off the round of the toe of the angle, thus making a foundation for the caulking. Then the toe is caulked and both ends of the angle clip are caulked to the plate as well as the heel of the back of the angle. In this way all four sides of the clip are made water-tight. If the ship fitter has laid off the rivets in the flange of the angle bar too far back from the toe, it is necessary to chip off enough material on that flange before caulking so that the material will not pry up when the caulking tool is being used.

When it is impossible to reach a joint with the straight

caulking tool because of an interference with some other portion of the ship a bent caulking tool is used in the same way as a straight one, but it can only be used for caulking, a splitting tool being necessary to prepare the material before caulking—(on straight caulking work the same tool may be used for both splitting and caulking).

In caulking butt joints the caulker works along this joint with the splitting tool, forcing a small portion of the material from each plate toward the opening between them and against the other plate. As the joint usually is a small opening the material from the two plates comes together during this splitting process and forms a wedge-shaped groove. (Fig. 26 *A*.)

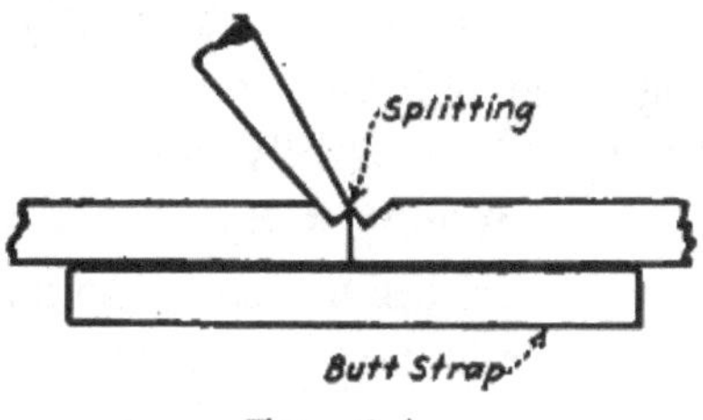

FIG. 26 *A*.

Then working with a roughing tool along the joint the two plates are hammered together tightly, taking care that none of the material used in forming the joint is allowed to project above the top side surface of the two plates. (Fig. 26 *B*.) The amount of material to be used from the edge

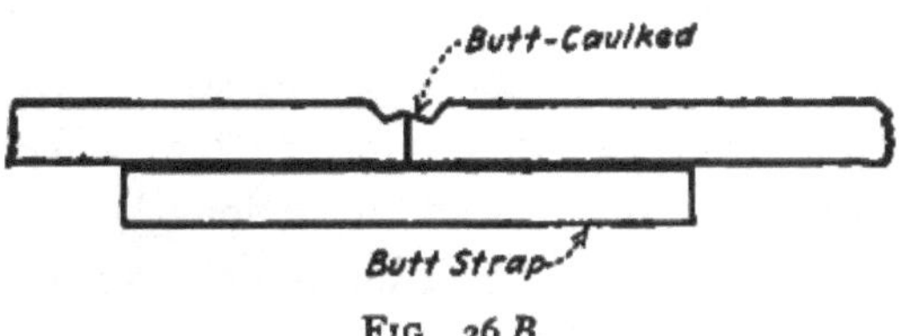

FIG. 26 *B*.

of each plate in making the wedge-shaped joint depends upon the size of the opening between them. With a narrow space a small amount would be used, while considerably more material would need to be dislodged in order to work over sufficient material to fill in a larger opening for forming a tight joint, care being taken at all times to keep a smooth surface.

Butt joint plates must be carefully fitted close together in order to produce a tightly caulked joint. The watertightness of a butt joint depends on the caulking, therefore it must be carefully done.

In caulking a butt joint the caulker must watch the line of the seam closely, so as to follow it. The caulking tool should not be allowed to slip first on one side, then on the other on the joint, because this will produce poor and bad-looking work. This is a type of joint uncommon on "merchant" vessels, rather on yachts and men-of-war.

When the opening between two plates is too wide to be caulked in the usual manner for a butt joint, a "shim" or "razor-back" wedge is used. It is a long, narrow, tapered plate with one edge brought to a "feather," with the full thickness along the back of the other edge. The length of this shim should be sufficient to fill the length of the opening into which it is driven with a hammer and material along the top plate is then split, the side movement of the metal which has been displaced holding the shim in place. The roughing tool is worked along this cut or split in order to broaden it out and throw the metal still tighter against the shim. The excess metal on the outside edge of the shim is chipped off fairly close to the plate. The roughing tool is used on this excess metal and the shim driven in against the plate on each side of it making a finish on the surface so that the joint presents an even appearance for the full length, and across what was the opening. Then the shim can be caulked in the same manner employed for any plate.

It often happens that a tapered liner is too short for its location and sometimes watertight angle bars are not quite long enough so that their ends do not touch, leaving an opening between the short splice angle and the face of the bulkhead. These openings are filled up by driving a four-sided steel plug in with a hammer and then holding it in place from jumping out, by forcing into it a little wedge-shaped piece of material from each of the two adjacent plates. Then the head of the plug is worked over the opening to prevent it from moving further into the hole. After this the plug can be caulked in the usual manner against the metal of the plates which surround it, making a watertight joint.

"Watertight staple angles" around floors are a source of trouble if the corners are not worked just right by the angle smith. If there are any openings between the back of the angle and the plate, it is necessary to fill in these places with steel plugs, as when tapered liners do not reach to the end of the next plate.

When driving these plugs into place care must be taken that they hold well and that one does not force out the other one near it. These are caulked in the usual manner, the same as described for the liner plug.

This applies especially to those corners which are "offset" or "crimped" back for about three and a half inches, to allow the continuous passage of an angle; as happens against the top and bottom angles of the centerline keelson. It is the endeavor of the anglesmith to carry this crimp just the correct length to suit the flange of the continuous angle. This is a difficult matter and it often happens that the small spaces between the heel of the floor angle and the bosom part of the flange of the continuous angle have to be plugged as the opening is too wide to fill in otherwise. (Fig. 27.)

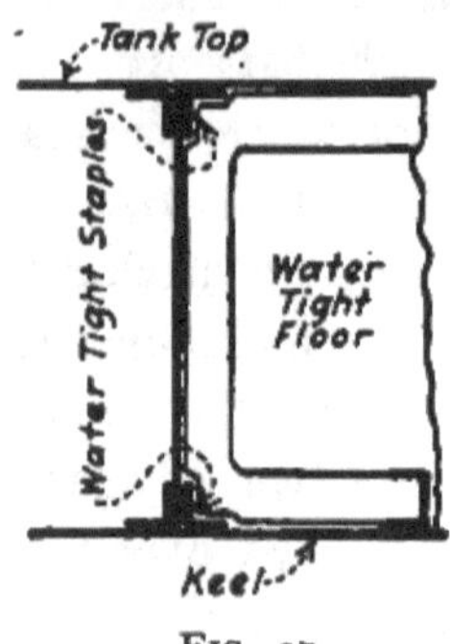

FIG. 27.

During caulking work, the workman often comes across "soft packing" of felt or canvas soaked in a preparation—sometimes linseed oil. This is used to make water tight some of the work which is very difficult for the caulker to get at.

This can often be a hindrance unless the caulker is very careful in his work, as the packing hardens after having been soaked in the oil or other preparation and its appearance is deceptive so that the caulker will work right along past the joint where this packing has been put in, without noticing it; especially if the packing happens to be of such a width as not to project beyond the joint sufficiently to be noticed.

The workman meeting this packing should take extra care with a testing knife to be sure that he has driven metal to metal and made a watertight joint. It is sometimes possible to detect the packing after the joint has been caulked by noticing the extra smooth, shiny appearance along a portion of the caulked seam. This smooth, shiny effect is due to the oil or other preparation in the packing.

A number of seams on a vessel are so placed that they must be caulked with the handle of the tool held in the left hand. The lap joints on the strakes of the starboard side of the ship are "left-hand" work. Skill in this requires practice.

Designers so place the parts inside of the ship that left-hand caulking is avoided, that is, the horizontal caulking edges face upward, and bulkheads caulked on forward surfaces have vertical edges facing to port, and those caulked on after surfaces have vertical edges facing to starboard.

The groove on shell plating should be $\frac{3}{16}$ in. wide. A narrow groove indicates a high shoulder. The depth of the groove is left to the judgment of the caulker. If plates are close together when caulking is started, the groove need not be deep. If they show open, as is often the case at the bilge, the groove must be deep to produce a tight seam.

When a plate is to be caulked its corner should be round rather than carried to a square edge as by so doing it is possible to force the metal for the caulking joint without prying up the material on the corner. All riveted joints are so designed that there is only enough material beyond the rivet sufficient to give ample strength to the joint, as an excessive amount of plate or angle beyond the rivet will tend to curl up from the plate below it when the edge of that plate or angle is being caulked.

In caulking oil-tight work, all lugs and angle bars are caulked all around it. In beveled angles the heels may be ⅛ in. away from the plate at places. It is necessary under such conditions to force soft packing into the opening, and then use the splitting and caulking tool.

Caulking a water-tight bulkhead should be done entirely from one side. If it is done partly on one side and partly on the other, water-tightness would not be secured.

If double frame angles are used at the bulkheads, only one is caulked; this is the after frame angle on forward bulkheads, and forward one on after bulkheads. The bulkhead should be caulked on the same side that the frame angle is caulked. Around the after peak and deep tank bulkheads, where soft packing or putty is placed at the frames, the caulker should be just as particular with his work as where packing is not used.

Where the overlapping deck plates are linered, these must be caulked. Where continuous parts, such as stringers and frames pierce water-tight bulkheads and tank tops, they are surrounded by "collars" of different forms. These collars often come to the ship already formed, are located by the erectors, and riveted up. They must fit closely around the part, and all should be on the same side of the bulkhead that is caulked.

The side of a bulkhead that is caulked should be open for inspection; that is, a tank should be caulked on the outside, whenever possible, because leaks during testing are much more easily located. Particular care should be observed in caulking, that the tanks in the double bottom are caulked on the open side. The oil storage tanks are caulked on both sides.

In testing a tank, if a leak is found that cannot be corrected by caulking, rivets must be driven out and renewed. To do this, the tank must be emptied and a second test made after the tank is filled again.

The shaft tunnel is caulked, and specifications may require testing this with water, the hose test being used. The decks are tested with the hose test. This test is described later. The ballast tanks of new vessels need very careful caulking.

Many times it is necessary to caulk the points of loosely driven flush rivets. When caulking these rivet points sometimes the metal on the end of the point is out too far and before it is caulked the excess stock is chipped off. The material is then hammered down with a "roughing" tool and it is ready for caulking. This is done by passing the "frenchman" around the edge of the rivet and making a slight groove between the material of the rivet and of the plate. A second or third trip is necessary to force the material of the rivet in toward the center and downward at the same time. As it is in a bevel opening, the rivet becomes tighter and is thus made water-tight. When doing this it is desirable to leave the metal of the plate untouched by the caulking tool, working only on the rivet point. (Fig. 28.)

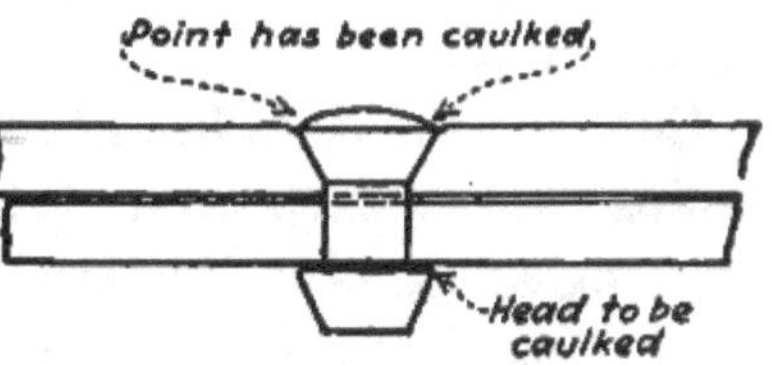

Fig. 28.

Many times a rivet head is not correctly formed in the rivet-making machine, and the resultant head is beveled outward toward the edge. If the rivet is not discarded by the riveting gang it will be driven up hard, the point properly hammered, and yet is very likely to leak and cause trouble, as water can get in under the head. The underside of the head of the rivet will touch the plate and bear hard on it, but a crack will show around the outer edge. If this crack is less than $\frac{1}{16}$ inch it can be caulked; but when it is greater it is difficult to obtain a good job, and it is best to cut out the rivet and drive a new one.

When caulking the head of the rivet a blunt, smooth tool should be used so as not to mar the surface. A "fuller" should not be used as it will form a ridge on the head and due entirely to the appearance, will often be condemned by an inspector.

Caulking the rivet head should be done around the side, taking care to keep some distance up from the edge, and working a little of the material of the head down toward the plate, thus filling in the opening between the under side of the head of the plate. This operation should be continued until the material is down hard against the plate and the crack has been entirely filled up. This is known as caulking "highhead" rivets (usually "panhead") and can be done to pass inspection if the workman is careful not to scar the material, as often a good, water-tight job will be thrown out if the inspector does not like the appearance of the head.

When caulking rivet heads through angle clips or under other conditions where one rivet is being tightened up and is liable to loosen the grip of another, it is necessary that the workman go over enough of these rivets to be sure that none of the neighboring ones have been loosened during the work.

The perfection of caulking work is tested by inspection of the groove formed by the tool. This groove should be of proper width and be deep, not just a slight mark. The edge of a thin-bladed knife is run along the seam; if it can be inserted at any point the caulking is poor, and must be done over. When inspecting in dark places, as at the bottom of the ship, a light is used so that the seam can be properly examined.

Chipping

The Chipper may be called upon to do "cutting" and "trimming." Chipping is taking off rough projections. Cutting is slicing off material along the edge to reduce the length, while trimming is smoothing and evening up an edge or surface.

Some of the sight edges as left by the bolters and riveters are not fair on bottom, on bilges, at sides, and on the decks. These unsightly places are corrected by chipping, before any caulking is done in the berth on which the caulker is working. The workman must be careful not to cut into the adjoining plate when chipping a sight edge. For this chipping a hexagonal shank chisel and the proper hammer for it are used.

To cut a plate in two or for any other deep cut along the edge, the caping chisel is used to make the first cut, going down into the material about $\frac{1}{8}$ of an inch, forming a groove for the full length of the cut. This is followed by the ripping tool which uses the path made by the caping tool as a guide and cuts much deeper.

The ripping tool should be kept in close against the plate

and in this way leave the edge smooth and clean both along the sides and at the bottom of the cut.

When two plates are butt-lap and joined to the plate in the next strake alongside of them, it is necessary to cut a "scarph" in the material in the end of the plate underneath. This reduces the "jump" occasioned by having three thicknesses of metal in line, and by cutting this tapered scarph the edge of the plate passing the butt lap can be run in a continuous straight line.

(NOTE.—If a scarph is not made a "tapered" liner is used.)

It is often necessary to cut the scarph along the landing of the plate where it is to lap over another. The width of this scarph is governed by the width of the landing as it is merely a bevel end near the edge of the plate usually between the second and third rivet hole, starting with the thickness of the plate and tapering to a point at the end. (Fig. 29.)

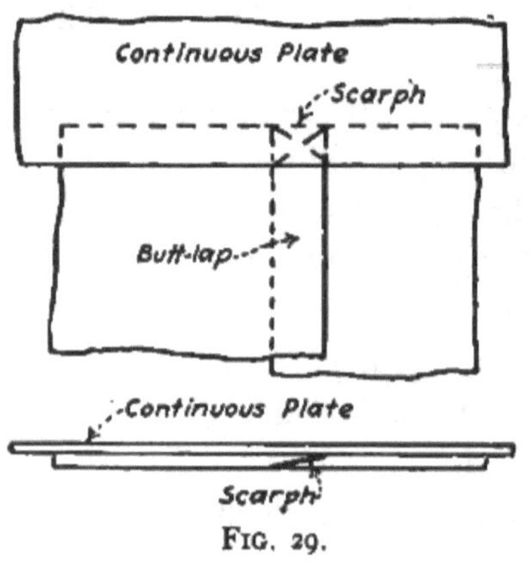

FIG. 29.

When chipping this scarph it is customary to cut grooves along its length with the caping chisel starting at the beginning of the scarph and working toward the end of the plate. The distance between these grooves should be governed by the width of the "side cutting" tool which is to be used next. After a series of these grooves has been cut, having a number sufficient to take in the whole of the width, then the side cutter is used in place of the caping chisel and the material is worked down smooth with the bottom of each cut. This operation is repeated with the caping chisel and side cutter until all the material which is to come out of the scarph has been chipped off, leaving a smooth surface. The width is equal to the amount of overlap of the adjacent plates along the side, and starts with a sharp bevel, the thickness of the plate, and runs to nothing at the end.

To chip a hole, varying from one-half inch to two inches in diameter, it is customary to lay off a circumference giving the location of the edge of the hole and then making four center punch marks at the ends of two diameters at right angles, beginning from one center punch mark and working across to the mark at the other end of the diameter. (Fig. 30.) The gouge starts with a slight cut going deeper in the center of the hole and coming to a light cut at the other edge. The same is done on the other diameter and this operation is repeated until the metal has become so weakened in the center of the hole that it is possible to put the gouge directly on end and break the metal through. The metal is then chipped directly down, working around from the center and enlarging this opening until the cut is made, out to the mark of the circumference, and finishing the hole to the desired diameter.

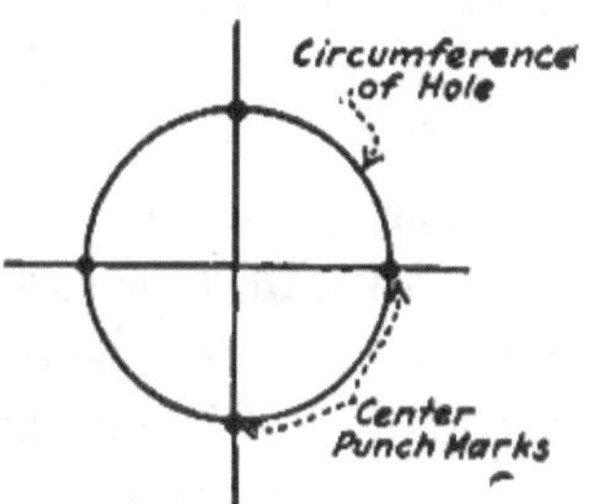

FIG. 30.

The chisel used should be sharp and ground at the proper angle. Chisels are furnished already ground, and if not properly ground should not be used.

All chisels and caulking tools are furnished the worker hardened and tempered. If the edge of the chisel does not cut when it is placed on the work, it is soft; the edge will be seen to turn back. Have a good stock of tools when starting on the job.

Sharp tools are important for good work. Every workman should know how to grind the points of his tools to the proper edge or bevel. As the pressure of the tool against the grinding material generates heat the tool should be occasionally dipped in water to keep the temperature down. If the tool is allowed to become overheated the temper would be drawn from the steel, making the tool useless.

During the chilly weather of winter it is a wise plan for the workman to be sure before he goes to work in the morning that his tools have not become chilled through during the night, as when in this condition they are liable to be very brittle. If there is a doubt in his mind, the workman should put these tools over a fire for a sufficient length of time merely to take off this chill, being careful not to heat them enough to draw the temper.

Duties of the Shipfitter

In this department there are divisions of Shipfitters, Regulators, and Linermen.

The Shipfitters are men thoroughly experienced in the shipbuilding work as regards the assembling of the ship structure. They must know every detail of construction about the ship and when any part of the ship comes in a little out of size, either too full or too scant, they must know how to make an alteration so that it can be properly adjusted. It is from this work that the name is derived, " Shipfitting."

It is also the duty of the shipfitters to lay out watertight staples around angle bars, fit intercostal shell angles to the decks, fit and install watertight doors, hatch covers, mooring rings, hawse pipes, rail and awning stanchions and all the other articles about the deck, which go to make up the deck fittings.

The shipfitters are on such work from the time it is started and are in charge of the assembling of the ship itself. The supervision and carrying out of detailed construction on the hull goes on until the vessel is launched and after that time until the ship is ready for delivery they are busy with the many small fittings required both inside and outside on the decks. The shipfitter, in addition to his special knowledge of shipwork, must be a good mechanic and accustomed to using his head as well as his hands.

The work is often divided up on the ship as follows: Two first-class fitters, with two or three helpers, work on frames. After the templets have been made in the Mold Loft the fitters get them, order the stock from the Steel Yard delivered to the Plate and Angle Shop where, after it has been furnaced and bent to shape, mark it up according to the templete, center punching all rivet holes and then surrounding the punch mark with a ring of white paint laid on with a round pipe or marker (often a small brass pipe); marking it up for shearing or the other operations yet to be done.

According to the type of ship, the fitters do as much assembly as possible before the material leaves the shop and often much of it can be riveted by the "bull machine" riveter (hydraulic) at once.

Another gang of two first-class fitters, with helpers, are busy with deck and Inner Bottom plating, laying out the material in the Plate Shop ready for shearing and punching according to the templates received from the Mold Loft.

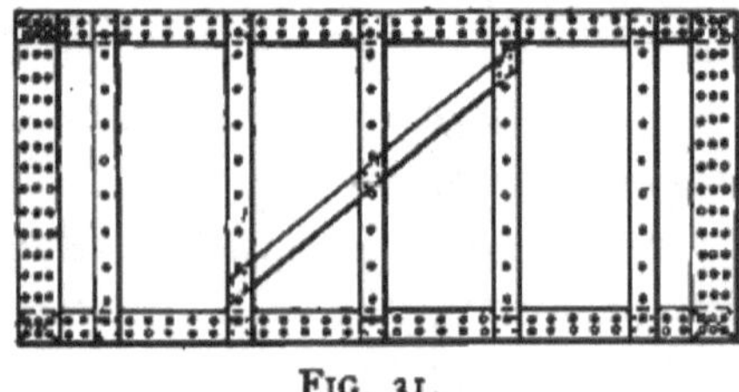

FIG. 31.

All plates on the "dead-flat" (perfectly flat, smooth surface) are easily laid off to a fairly good fit, but others, at ends of the ship where they meet the turn of the side plating, are usually lifted "from work" (on the ship).

A template for a plate as shown in Fig. 31 would be made in the following manner. Suppose it is for an outside deck plate. The fitter who lifts the plating must be sure that the material in the ship is already "fair." The edge of the plate is laid off, for the landing, measuring from the outer line of rivet holes in the under plate; about 1¼ inches for a ¾-inch rivet and about 1½ inch for a ⅞-inch rivet. The outer edge of the wood template strip is laid along these marks, shaping it with a knife if necessary. This is repeated for the other side and the two ends and narrow strips carried across the beams and all the pieces of wood tacked together. If the template is a long one it is well to put on an extra wide diagonal strip to prevent it from collapsing.

All rivet holes are next marked by tracing them with a pencil through the holes in the plates which have already been punched. The template should be marked with all information as to hull, location, size of holes, countersunk or not, etc., and then sent to the Plate Shop.

During the process of forming the template the separate pieces of wood are held in place on the steel plates by means of steel clips made of round bars about one-quarter inch diameter and bent into the form of a letter U, with the ends coming close together and bent outward at the tip. (Fig. 32.)

FIG. 32.

Here the template after being bored out for all rivet holes as marked on the under side is laid on the plate, which has already been ordered out from the Steel Yard and is laying ready for the "fitter" so that all rivet holes can be center punched, painted, the outside edge of the plate marked, and symbols painted on it to make it ready for the workmen of the Plate Shop.

Another gang of fitters, usually two working together, with a gang of helpers, sometimes as many as fifteen, are busy on the shell plating. These men "lift" all plates of unusual shape, such as the furnaced plates forward on the lower end of the stem, aft on the stern frame, boss, oxter (tuck) plates, etc. . . The "outer strake" of the plating is usually lifted from the ship by this gang. They come after the other plating has been "regulated" and bolted and that portion of the ship is "fair" and make templates for each plate; using them for both port and starboard sides wherever possible. The work is done in a manner similar to that just described for deck plates, although more clips must be used as some of the plates at the "shoulders" of the vessel have "sny" (curvature both across and lengthwise of the plate and the battens must be held close up to the work).

Work on bulkheads is done by two first-class fitters with helpers. Splice bars (often called "bosompiece") and some staples are lifted from work by this gang but most of the bulkhead material is laid off in the Mold Loft.

Two other first-class fitters with helpers are busy with splice bars for stringers, girders and longitudinals which are lifted on the ship. The plating and angles for these members are laid off in the Mold Loft.

There is considerable miscellaneous work on butt-straps, liners, doors, hatches, deck fittings, manholes, etc., which requires a few first-class or second-class fitters with helpers.

This description of the method used in getting material ready for assembly on the shipways applies to the majority of shipyards and is more or less standard throughout the country.

The Regulators follow directly behind the Erectors and move the plating a fraction of an inch one way or another in order to take up any discrepancies in one plate by allowing a little shifting in a number of plates.

The Regulators use drift pins and heavy mauls in their work and after the plate is in position they bolt it in place, ready for the gang of Bolters-up who will thoroughly bolt it tight for the Riveters. When the rivet hole does not come fair, it is customary to use a smaller size bolt than the size of the hole. In this way a $\frac{3}{4}$-in. bolt would be used in a $\frac{7}{8}$-in. rivet hole which was slightly unfair. The regulating of a ship requires practice and the work of the Reamers is considerably less if the regulating men have done their work thoroughly.

The Linermen follow up the Erectors, if the system of "in and out" plating is being used. In this system every other strake of plating is "raised" and requires a distance piece between the inside of the plate and the framing. These liners or distance pieces vary in size and thickness, according to the location. They are usually divided into "parallel" and "tapered" liners. The parallel liners are about the width of the flange of the angle which they touch and their thickness is governed according to the thickness of the inner strake with which they are in line.

The liners are punched for rivet holes, according to the spacing of the holes in the plates. The tapered liners are fitted under seams where a butt lap is used and are made to fill the triangular space which is left open.

The width of the liner is governed by the width of the seam (the lap on the edge of the plate) and is generally long enough to take at least two rivets, which are sufficient to hold it in place. These liners are usually delivered to the ship with a symbol painted on them, indicating their location and it is the duty of the Linermen to distribute these according to the marking and see that they are in place ahead of the Bolting-up gang, so that when the Riveters are ready for their work there will be no delay.

When laying out tapered liners it is important that the distance between the center of the rivets and the outer edge of the liner be accurate as otherwise it will have to be chipped off if out too far and the plate will have to be chipped back to it if it is not out far enough.

Parallel liners against an outside strake of the bottom shell plating are usually cut about two inches short from the edge of the next inside strake to allow for drainage of the Inner Bottom. Those liners against the under side of outside strakes of the Tank Top are also cut short about the same amount but in this case it is for passage of the air as it rushes along the top of the tank when it is being either filled or emptied.

On the so-called "Fabricated" ships (which should really be called "manufactured" ships) which were constructed in quantities during the stress occasioned by the World War, an entirely different method of preparing the material was used.

Molds were prepared in the Mold Lofts of the yards which were to construct vessels in wholesale quantities and then shipped to all parts of the United States and Canada. Plate shops of all kinds were pressed into service and plates and shapes turned out in quantities of ten, twenty or twenty-five, as the case might be. Certain shops made certain parts of the ship but the plates or shapes (angles and channels) were formed as required, punched for rivet holes and countersunk where necessary. That material came into the shipyards all ready for assembly on the ship.

In such cases all the floors came in riveted complete with frame and floor bars all in place. The water-tight floors came in all caulked ready for installing on board. At the Hog Island Yard the vertical keelson came in with two plates riveted together at the butt-lap and all four angles riveted on, two on top and two at the bottom, with enough overhang to allow for "shift of butts" of the ends of the angles.

For these boats, 401 ft. overall length and 54 ft. (molded) beam, of 7500 tons dead weight carrying capacity, about 95 per cent of all the hull steel came to the shipyard all fabricated and ready to put together. Because great care had been taken with the work in the Mold Loft, followed by careful inspection work at the shops; the rivet holes of the frames and shell plating in the "molded" (curved) ends of the ships fitted almost to a perfect match. Needless to say that portion of the vessel amidships, where the sides and bottom were flat surface, was well matched and caused no difficulty in assembling. The work done on these boats was a surprise to many "old-time" shipbuilders because this matching of the rivet holes was far superior to that seen in many vessels at other yards where the system previously described is in vogue, where a portion of the shell would be taken from the Mold Loft and the "outer" strakes lifted from the actual conditions on the ship.

The vessels at Hog Island were "in-and-out" plating, and all the outer plates as well as the inner ones were laid off in the Mold Loft and made just as good a fit when they arrived at the ship. Because of this it was possible to erect and bolt loosely in place outer strakes which would be later pried up and the inner ones slid under and into position whenever the irregularity in the arrival of the material at the yard meant that procedure or a hold-up in the work.

Keel plates at the forward and after ends of the ship, the boss and oxter (tuck) plates, the plates on the shoulder of the ship, just forward and aft of the dead flat (parallel portion), part of the bilge strake, were those which came to the yard without rivet holes punched in them.

This shows what it is possible to do in this new style of building. Thirty-eight mills and eighty-eight fabricating shops (many of them boiler and bridge shops) were employed in this work and they extended from Montreal to Virginia and as far West as Kansas City. Yet the different parts fitted almost perfectly.

Duties of the Ship Carpenters

The Ship Carpenter does all the rough woodwork about a ship, including the building and truing of shipways, setting of shores, placing the staging and setting up backing for Riveters. His tools are much the same as those used by house carpenters, namely saws, hammers, levels, nails etc. The adz, an axe with a thin curved blade set at right angles to the handle, is used by ship carpenters because it is especially useful for chipping and slicing away the surface of wood.

As the work of ship carpenters on wooden vessels is one of the main features of their building, and there is a great deal required of them that does not go into the work of ship carpenters in steel shipyards, their work will not be considered here.

The permanent Shipways are built by a contractor who places the piling for the foundations and builds the cross-bents and decking.

The carpenter's responsibilitystarts with the keel blocks, and these are built, plumbed and the top block beveled to suit the inclination of the ship as given by the engineers. They strike the center line of the boat along these blocks after the top has been beveled and they also locate the "dead flat" or midship mark.

In those shipyards where it is customary to erect a monument or small upright joist to serve as a positive spot for the location for the first keel plates to be laid, the work is given to the ship carpenters. They are also responsible for the horizontal spalls built in line with the keel to receive the bottom shell plating when this is laid, and they later replace these spalls with the shores for greater strength.

Ship carpenters fit diagonal bracing against the after side of the keel blocks to prevent any motion of the ship down hill, due to the inclination of the keel as the ship is being built.

After the hull is well started, the work is divided up between a gang on the outside of the ship and another gang on the inside. This is necessary, as the need for staging must come under the eye of the foreman and where work is progressing rapidly he must be able to see where he can prepare in advance of the steel workers.

Staging work and backing does not require a great deal of skill, but men must be handy with woodworking tools. Some staging work is nothing more than planking on a couple of horses. On other occasions, it may demand bolting uprights to various portions of the ships and putting in thwarts on which to lay stage planking.

Carpenters who work on staging are called "stage builders." This staging consists of outside and inside staging for the hull of the vessel. The outside staging is laid ta various heights to suit the Riveters and other workmen who are to work along the edges of the shell plating, and when it becomes necessary for any of this staging to be changed in height for the benefit of the other workmen, the ship carpenters attend to it. This outside staging includes the work along the side of the ship and at the bow and stern. Around the bow arranging upright and planking for the benefit of the men is fairly easy, but around the stern it is more difficult, due to the fact that it is necessary for the workmen to be in close to the ship and the over-hang of the counter of the vessel requires an elaborate system of uprights and braces in order to hold the staging and planking in close to the vessel, and at the proper height.

Just before the time for launching the ship carpenters remove all the staging from around the stern in order to give a clear passage for the boat during its travel down the shipways. The staging along the side of the shipway is often left standing and used on one vessel after another.

Inside staging in some parts of the boat is a simple matter, as for instance along the decks, where there is a height of only seven or eight feet, all that is necessary are horses with planks laid across them. This is good for the Holder-on under the deck above or for transverse bulkheads, for both the Riveter and Holder-on.

In the cargo holds the staging must be much more elaborate, and it becomes necessary to put in quite heavy uprights to take the staging along the side shell and on the cargo bulkheads.

In some shipyards special "horses" are made for the bulkhead work with an arrangement so that the stage planking can be easily adjusted for different heights to suit the work. Built this way the staging can be used many times with out dismantling and by means of an over-head crane can be lifted bodily into the cargo holds and out again, as desired. Using horses avoids a lot of the necessary cutting and sawing with the consequent loss of material which obtains when staging is built for a particular place and then dismantled. The staging around the Cargo Hatch coamings is more or less elaborate in order to enable the Holder-on to work to advantage without danger of an accident. Many of the yards fitting a "back rail" have this staging the same as for outside staging.

The ladders having access to the different parts of the ship during the construction are built by the carpenters, who must always be ahead of the job in order to have these built and ready for use as soon as the steel is high enough to require them. Ladders wide enough for two or three men to pass at the same time arc provided reaching from the bottom of each Cargo Hold to the Upper Deck.

Many times it is necessary to put in temporary shoring under the deck beams until the regular steel stanchions are in place. This work is done by the ship carpenters who follow along with the erectors, watching carefully to help support anything, as needed.

In some shipyards where they have facilities for lifting heavy weights, bulkheads, shaft alleys deck houses, etc., are built on the ground at the head of the shipways.

Because of local strain, as these are being moved from the ground to the ship by cranes, it is necessary for the ship carpenters to brace these structures temporarily with heavy joists thoroughly bolted thus preventing any distortion.

The ship carpenters are on the lookout at all times to forestall the needs of the steel workers in the matter of additional staging, removing old staging, or altering it.

In the case where two timbers come together, it is better to bolt than to spike them, because the timber is in better condition for use at some future time. There are always plenty of bolts and angles about the Shipyard which may be used for connecting staging timber.

Many times a "jam riveter" must be used and the ship carpenters will be called upon to arrange the backing in heavy planks at proper distance from the plate to accommodate this tool.

As the boat increases in weight, it is customary in some yards to build cribbing under the bottom of the ship near the bilge, there being about three or four cribs on each side of the center line. This cribbing is a stronger support and more reliable than the ordinary shoring.

All preparations for launching are made by the ship carpenters as they place the ground ways, align them carefully and then move the launching "cradle" in on top.

The "packing" of wood planks which is fitted between the top of the sliding ways and the bottom of the ship is careful work and oftentimes the use of an adz is required as the "packing" must be shaped to suit the shell plating, allowing for the change in shape in way of the butts of the plating.

The only time when anyone really might envy the ship carpenters is when they are at work on the "saw-off" plank just as the ship is being launched, for theirs is the honor of doing the last work on the boat before she takes the water.

CHAPTER III

Shipyard Tools

Following is a list of tools used by **Bolters, Erectors, Shipfitters, Linermen, Regulators:**

- $\frac{3}{4}$-in. and $\frac{7}{8}$-in. Spud wrenches with offset,
- $\frac{5}{8}$-in. wrenches with offset,
- Spud bars,
- 1-in. pipe handle,
- Testing knives,
- Center punches,
- 2-lb. and 8-lb. Mauls,
- $\frac{3}{4}$-in. and $\frac{7}{8}$-in. Drift pins,
- Soapstone,
- Chalk,
- Yellow crayons,
- Turnbuckles,
- Steamboat ratchet,
- Heel wedges,
- Straight wedges,
- " C " clamps.

Reamers and **Drillers** use these tools:

- No. 2 and No. 3 Air drill machines,
- No. 9 Corner drill machines,
- Ratchets,
- 2-in., 4-in. and 12-in. Extension sockets, 3–3,
- Sleeve sockets, 3–4,
- 4-in. Sockets, 4–4, 2–4, 2–3, 1–3,
- $1\frac{5}{8}$-in. and $1\frac{7}{8}$-in. Countersinks,
- $\frac{15}{16}$-in., $\frac{13}{16}$-in. and $\frac{11}{16}$-in. Reamers,
- $\frac{15}{16}$-in., $\frac{7}{8}$-in., $\frac{3}{4}$-in. and $\frac{27}{32}$-in. Drills,
- Taps,
- Grease guns,
- Oil cans, 1 pt.,

Drift pins, Nos. 1, 2, 3,
Old men,
Eye-safe goggles,
Air hose,
1-in, Pipe handles,
" C " clamps.

The following tools are used by **Riveters, Holders-on, Heaters** and **Passers:**

No. 60 Guns,
No. 2 Chippers,
Holder-on machines,
Jam riveter,
$\frac{3}{4}$-in. and $\frac{7}{8}$-in. Snaps,
Concave dies, $1\frac{7}{16}$ in.,
Flush dies, $1\frac{3}{8}$ in.,
$\frac{7}{8}$-in. and $\frac{3}{4}$-in. Holder-on dies,
Heating tongs,
Passing tongs,
Catching cans,
" Y " Leader hose,
Hot chisels,
Eye-safe goggles,
Pint copperized oil cans,
Backing out punches,
Hand riveting hammers,
12-lb. Mauls,
Electric extension lights,
1-in. Pipe handles,
Gooseneck dolly bars,
Straight dolly bars,
No. 0 Dies,
No. 0 Countersunk dies and socket,
Rivet forge,
Testing knives,
Rivet testing hammer,
Red lead gun,

These tools are used by **Chippers** and **Caulkers:**

No. 1 and No. 2 Chipping guns,
Bobbing chisels,
Cape chisels,
Side cutters,
Gougers,
Roughing — straight, bent, fine,
Rivet caulkers—bent,
Rivet caulkers (Frenchman),
Fullers—straight and bent,

Caulkers—round, straight, bent,
Rippers,
Caulkers' dies,
Oil cans,
Eye-safe goggles,
Hand chipping tools,
Cold chisels,
Caulking chisel, straight, bent, round, fine, coarse,
Roughing chisels, straight—bent,
Diamond points,
Straight rivet caulkers,
Cape chisels,
Fullers—straight, bent.

Spud Wrench (Fig. 33): This tool is used by erectors or plate hangers when putting the plates together. The pointed end is used to direct the plate so that one rivet hole runs center over the one below it and the other end of the wrench is used for setting up bolts and nuts on the plates.

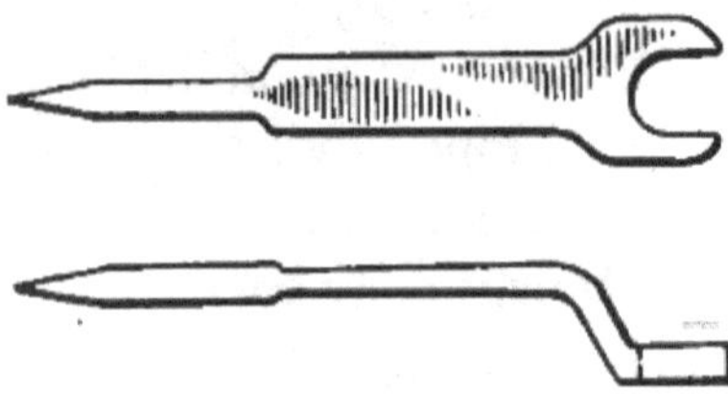

Fig. 33.—Spud Wrench.

Spud Bars (Fig. 34): These vary in length from 18 to 36 in. and are used in handling the plates when regulating them into the exact location required. One end is round and the other end tapers to a chisel point.

Fig. 34.—Spud Bar.

Rivet Testing Knives (Fig. 35): This knife is used by the rivet testers when examining the work to be sure that the joints between the plates are tight, without opening between them.

Fig. 35.—Rivet Testing Knife.

Figs. 36 and 37. These are the rivet testing knives which have been ground down to a shape as desired by

FIG. 36.

FIG. 37.

Rivet Testing Knives.

the workers. They are shown to give an idea of the different types which the men prefer.

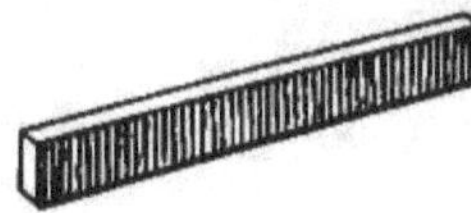

FIG. 38.—Soap Stone Marker.

Soap Stone Marker (Fig. 38): These are used as pencils by the steel worker as they give a clear line on the metal and can be sharpened down to a fine edge if required.

Mauls (Fig. 39): These are used for any heavy hammering about the ship, as by the regulating and bolting-up gangs.

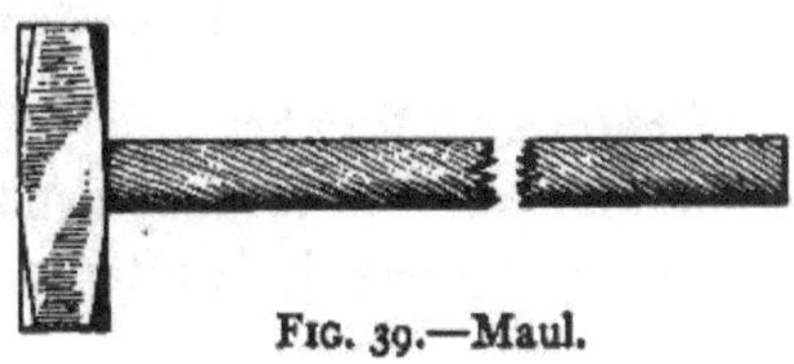

FIG. 39.—Maul.

Mauls (Fig. 40): This maul is used for backing-up when riveting and is used in place of the ordinary air tool for holding-on or in the place of a dolly bar.

FIG. 40.—Maul.

Drift Pin (Fig. 41): This is used for centering the plates when regulating them into the exact location required. When a plate is laid and the rivet holes are not exactly in line or "fair," the drift pin is hammered in, and, being in this tapered shape "round," it draws the two plates in line so that the centers of the rivet holes come one over the other, thus forming a "fair" hole down between the two plates.

FIG. 41.—Drift Pin.

Turnbuckle (Fig. 42): This is often used when drawing plates into position.

FIG. 42.—Turnbuckle.

Steamboat Ratchet (Fig. 43): This is used for work somewhat similar to the turnbuckle, but because of the hooks on the end it is often preferred. By swinging the ratchet handle the casting revolves, the right- and left-hand threads inside draw the two end links together (both ends are alike).

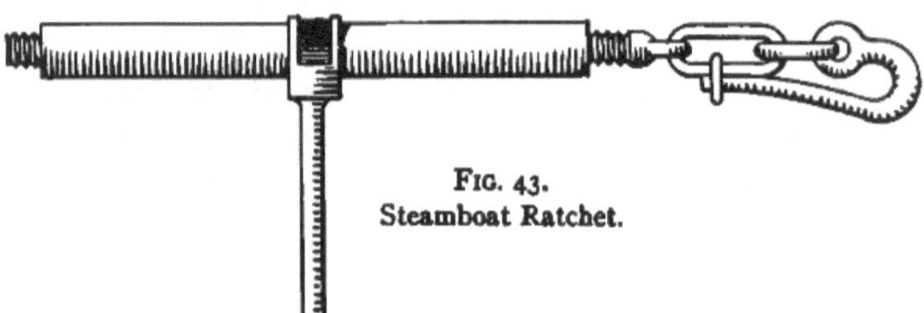

FIG. 43.
Steamboat Ratchet.

Straight Wedge (Fig. 44): This is used when regulating plates or when it is desired to raise the edge of a plate when fitting in liners or for some other reason.

FIG. 44.—Straight Wedge.

Heel Wedges (Fig. 45): These wedges are used in the same kind of work as the straight wedge and have the advantage of the heel on top, which is used for hammering against and backing-out the wedge.

FIG. 45.—Heel Wedge.

Air Drilling Machine (Fig. 46): This machine is used for drilling and reaming. It is operated by means of two handles, air hose being attached to the valve which

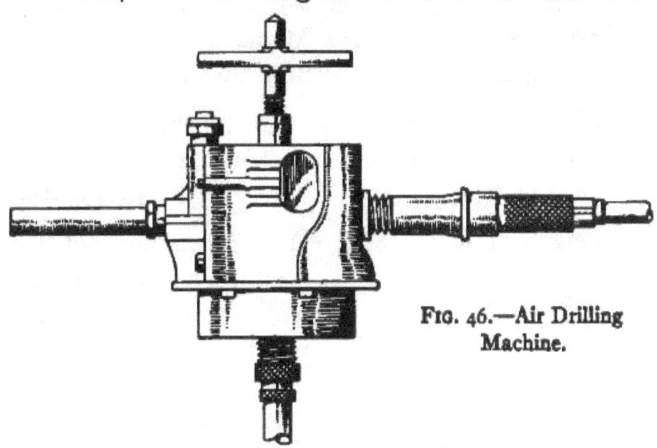

FIG. 46.—Air Drilling Machine.

is on one of the handles. The screw extension at the top is adjustable to hold the machine up to the work and is extended as the drill enters the hole which it is boring.

Corner Drilling Machine (Fig. 47): This machine is used in corners or any other location where there is small

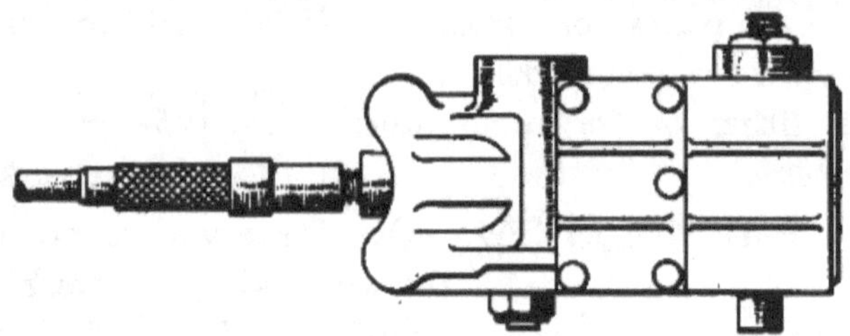

Fig. 47.—Corner Drilling Machine.

space. The screw at the top is set up with a lever which turns through a short arc. The valve is used as a handle on the back of the machine.

Wrench (Fig. 48): This is used for setting up the screw in the top of the Corner Drilling Machine, to raise the screw as the drill advances into the material.

Fig. 48.—Wrench.

Steel Socket (Fig. 49): This is used in the Corner Drilling Machine and takes the drill.

Fig. 49.—Steel Socket.

Hand Ratchet (Fig. 50): This is the old fashioned ratchet which has been used for drilling holes for many years. It is used in conjunction with the "old man" (Fig. 59), one hand moving the handle and the other tightening the back screw to maintain the pressure on the drill.

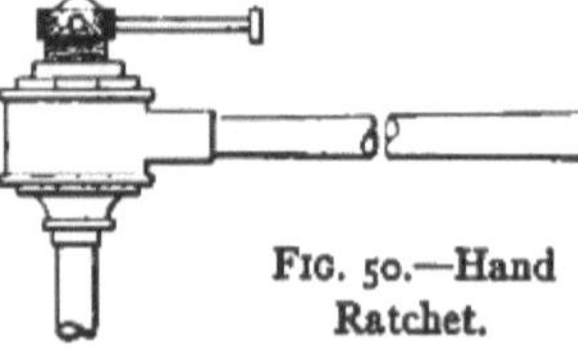

Fig. 50.—Hand Ratchet.

Countersink Drill (Fig. 51): This is used for countersinking any of the rivet holes as required.

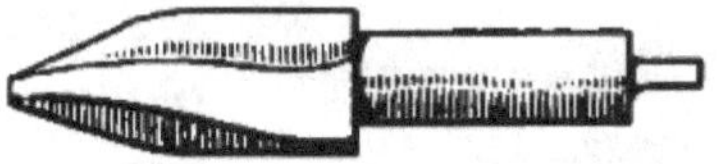

Fig. 51.—Countersink Drill.

Reamers (Fig. 52): The reaming tool is used in the air drill and comes in various sizes to suit the different size rivet holes. It makes the hole uniform, so that the rivet will be able to slide through easily before it is riveted up.

Fig. 52.—Reamer.

Drills (Fig. 53): These are ordinary machine tool drills used in the air drilling machine and are used the same as for any work where a straight hole is required.

Fig. 53.—Drill.

Taps (Fig. 54): These are screw thread cutting tools which are turned down into a hole after it has been drilled, in order to cut a thread for a stud or bolt.

Fig. 54.—Taps.

Tap Wrench (Fig. 55): This wrench is used for turning the tap when threading a drilled hole.

Fig. 55.—Tap Wrench.

Grease Gun (Fig. 56): This is a squirt can, or gun, used to lubricate the different air machines when they are not in use. By means of the plunger, grease is forced out of the gun and drops onto the bearings of the machine.

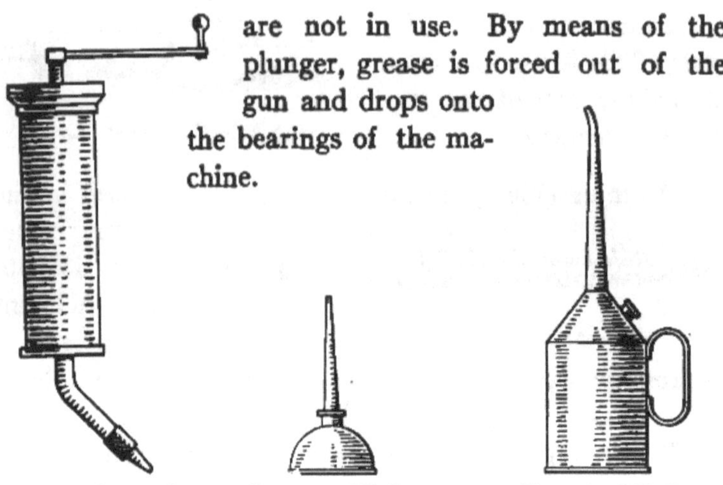

FIG. 56.—Grease Gun. FIG. 57.—Oil Can. FIG. 58.—Oil Can.

Oil Cans (Fig. 57): These are used for "oil" or "soup" (soap and water) for use of the Drillers in lubricating the work when drilling. Fig. 58 is another type of can used for the same purpose as Fig. 57.

"Old Man" (Fig. 59): This is a rig with a standpipe and base in one piece, a portable arm which can be raised and lowered or swung, to any angle, as desired. It is used when drilling, the base being on the material and the arm swung around until it is over the drill for which it forms a support and takes the thrust when the drill enters the material.

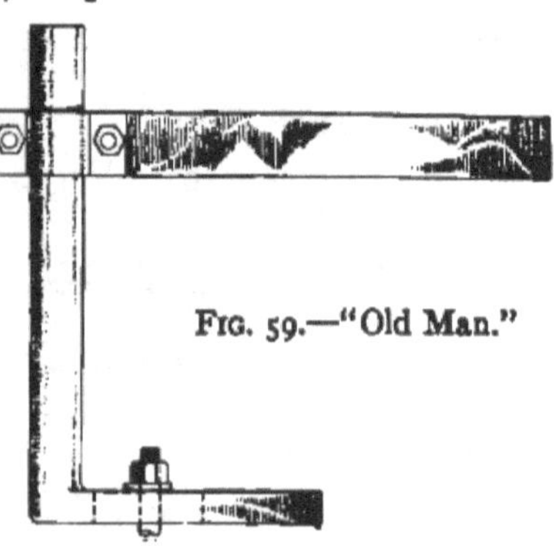

FIG. 59.—"Old Man."

Eye Goggles (Fig. 60): These are used by the Chippers and Caulkers or other men who are doing similar work as a protection against injury to the eyes. There are a number of designs, but any ordinary goggle is all right for this kind of work.

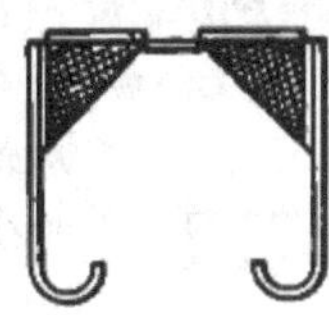

Fig. 60.
Eye Goggles.

"C" Clamp (Fig. 61): This is used by the Erectors or Plate Hangers and Regulators. It is designed to fasten the edges of two plates together and can be set up tight by means of a wrench on the head of the bolt.

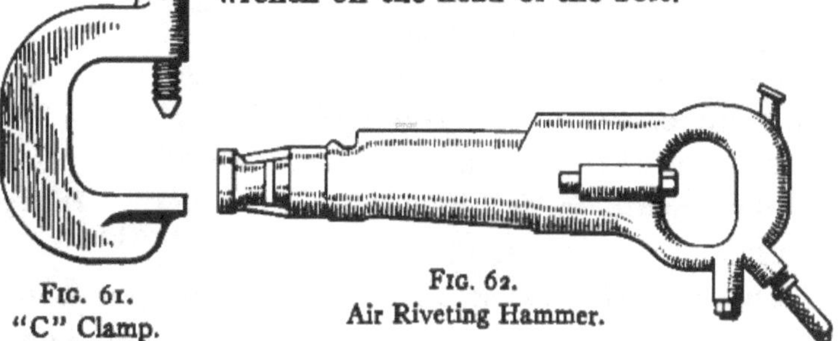

Fig. 61.
"C" Clamp.

Fig. 62.
Air Riveting Hammer.

Air Riveting Hammer (Fig. 62): This is used for driving rivets, the quick action of the hammer forces the hot metal into the rivet hole and forms the point of the rivet before it has much time to cool off. It is commonly spoken of as an "Air Gun." Catalogue numbers are used in reference when speaking of the size of hammer. Running from No. 40 to No. 90, the size most commonly used being No. 50 for $\frac{3}{4}$-in. rivets, and No. 60 for $\frac{7}{8}$-in. rivets. The hammer is operated by a strong air pressure, released by the trigger (shown on top of the handle).

Plunger (Fig. 63): This is a steel plug which is a "go-between" for the hammer and the rivet die. It works loose in the air gun and drives on the end of the die. This plunger is loose and should be carefully handled by the inexperienced man as it can be shot out of the gun with sufficient force to badly wound another workman, if it should hit him. To be sure of it, many of the riveters carry it in their pocket, when not in use.

Fig. 63. Plunger.

Rivet Die (Fig. 64): The rivet die is also a loose member of the Air Riveting Hammer family, The stem of the die coming in contact with the plunger in the hammer, transmits the blow to the rivet and forms the point of the rivet according to the shape of the die. The die shown in this figure is for a "Button Point."

Fig. 64. Fig. 65.
Rivet Dies.

Fig. 65. This die is similar to Fig. 64, but is used for a "Countersunk Point," often called "Flush."

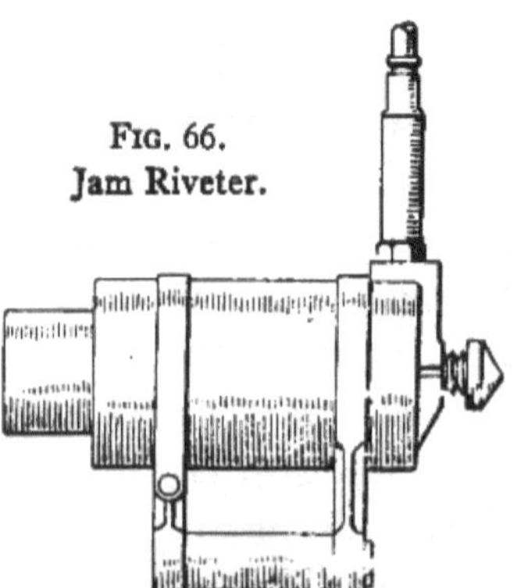

Fig. 66.
Jam Riveter.

Jam Riveter (Fig. 66): This riveting machine is used where it is possible to place the butt end against another part of the structure. This takes the bearing strain off the riveter and he can do considerably more work under these conditions, but it is seldom that this type of riveting machine hammer can be used on the general run of shipwork, because of the fact that it needs a back-up.

Air Holding-on Machine (Fig. 67): This is used by a holder-on who backs up the rivet on the other end from the riveter. The extension pipe with a set in the end fits directly against another part of the Hull structure and takes the force of the blow. Air comes in through the side and is under the control of the operator, and forms a cushion for the blow. The die in the end varies according to the kind of rivet being used.

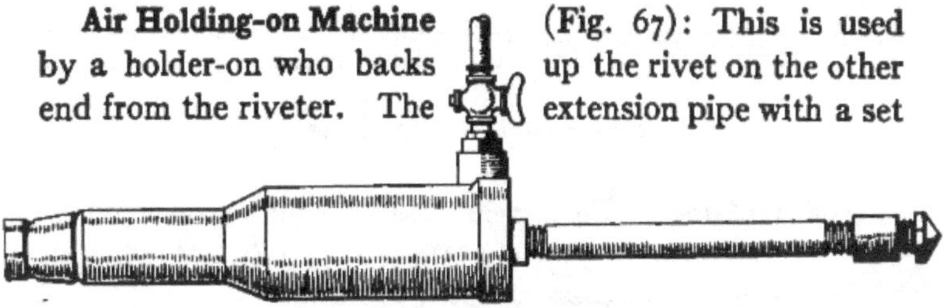

FIG. 67.—Air Holding-on Machine.

Bevel Holder-on Die (Fig. 68): These are fitted up either pan head, concave, or flush, according to the type of rivet and are used in odd corners where it is impossible to handle an ordinary tool which is square to the surface (shown "pan head").

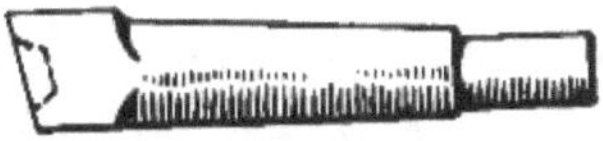

FIG. 68.

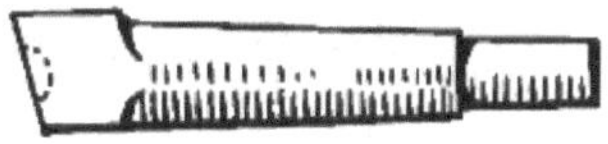

FIG. 69.
Bevel Holder-on Dies.

Fig. 69. This tool is similar to Fig. 68, but is faced for a "Button Head" rivet.

Rivet Heating Tongs (Fig. 70): These are tongs which are used by the Heater Boy at the rivet forge putting rivets in and taking them out of the fire.

Fig. 70.—Rivet Heating Tongs.

Passing Tongs (Fig. 71): These tongs are used by the Passer Boy in relaying the rivet and when putting it in place in the rivet hole.

Fig. 71.—Passing Tongs.

Catching Cans (Fig. 72): These are used by the Passer Boy to catch the hot rivets as they are thrown to him by the Heater Boy near the forge fire.

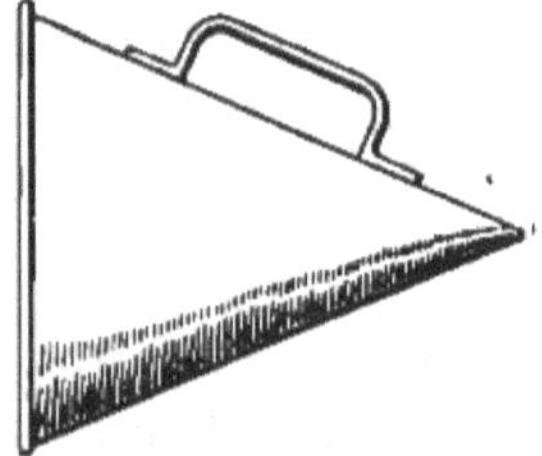

Fig. 72.—Catching Can.

"Y" Leader Hose (Fig. 73): Air hose for the riveter is generally branched for two leads. The main hose from the manifold is $\frac{3}{4}$ in. (rubber) and this is fitted to a "Y" which has a $\frac{3}{4}$-in. and a $\frac{1}{2}$-in. outlet. The larger one is carried to the riveting hammer and the smaller is

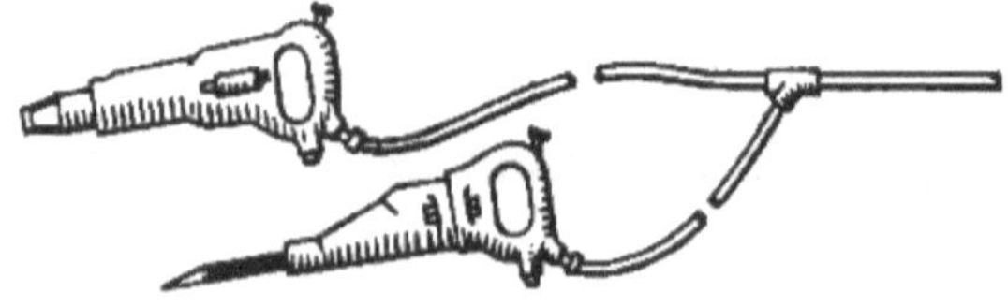

Fig. 73.—"Y" Leader Hose.

carried to the chipping hammer, This branch is about 8 ft. long, giving sufficient length for working without having too much hose to drag around.

Air Chipping (or Caulking) Hammer (Fig. 74): This is similar to the Air Riveting Hammer except that it gives a lighter blow and is used for a different purpose. When any of the plates have a ragged edge or a part must be chipped off, this hammer is used. It is also used with caulking tools when that work is done after the riveting has been finished. The riveters use this hammer as part of their outfit when driving flush rivets, the stock of which is too long. After forming most of the point of the flush rivet, the excess (called the "rag") is chipped off while still hot and the remainder of the material finished by the riveting hammer, before it has had time to cool off until hard.

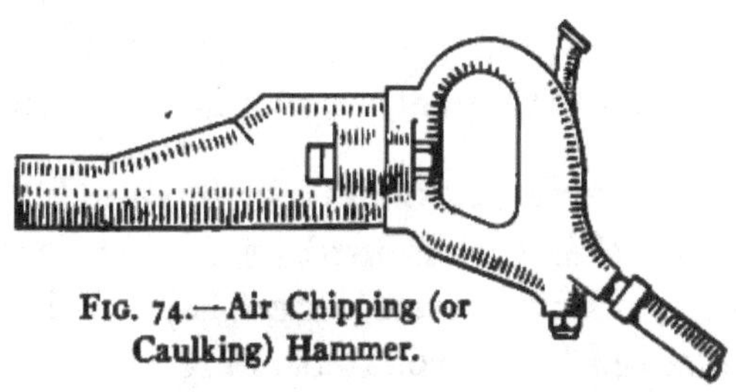

Fig. 74.—Air Chipping (or Caulking) Hammer.

Hot Chisel (Fig. 75): This is used in an air-gun to cut off the excess on the point of a flush rivet. When driving a flush rivet, the length of the rivet must be exactly right and the riveter must be on the safe side, so when the rivets are not just the length required, he will use those which are a little too long and then cut off the excess, called the "rag."

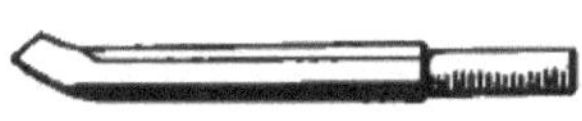

Fig. 75.—Hot Chisel.

Backing-out Punch (Fig. 76): This tool is used by riveters when it is found necessary to remove a rivet. The head of the rivet is cut off or burnt off and then the rivet is driven out through the hole by means of this punch, and a hand hammer.

FIG. 76.—Backing-out Punch.

Backing-out Countersunk Die (Fig. 77): This tool is used for backing-out countersunk rivets, the shape being adapted to the countersinking.

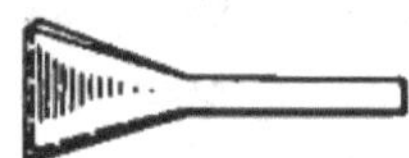

FIG. 77.—Backing-out Countersunk Die.

Die Socket (Fig. 78): This socket is used in conjunction with the backing-out die, Fig. 77.

FIG. 78.—Die Socket.

Tomahawk (Fig. 79): This is a blunt-nosed backing-out punch and is used sometimes in place of the sharper-nosed punch, as shown in Fig. 76.

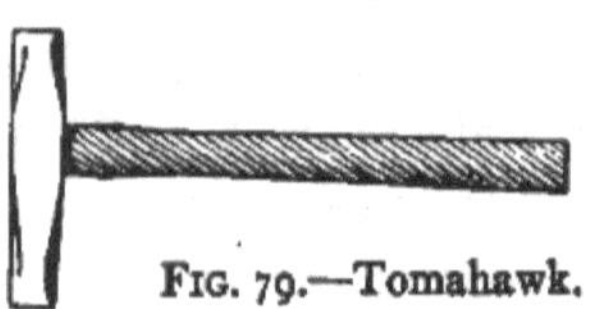

FIG. 79.—Tomahawk.

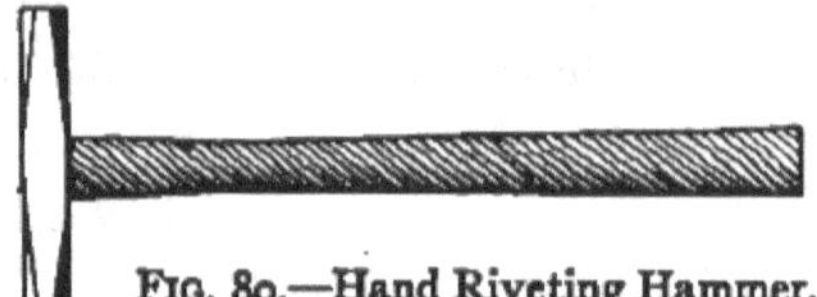

FIG. 80.—Hand Riveting Hammer.

Hand Riveting Hammer (Fig. 80): This hammer is used for riveting by hand when the work is of such a nature as to require handwork rather than the use of an air gun.

Gooseneck Dolly Bar (Fig. 81): This is used for hand work when holding-on for the riveter, in corners or some

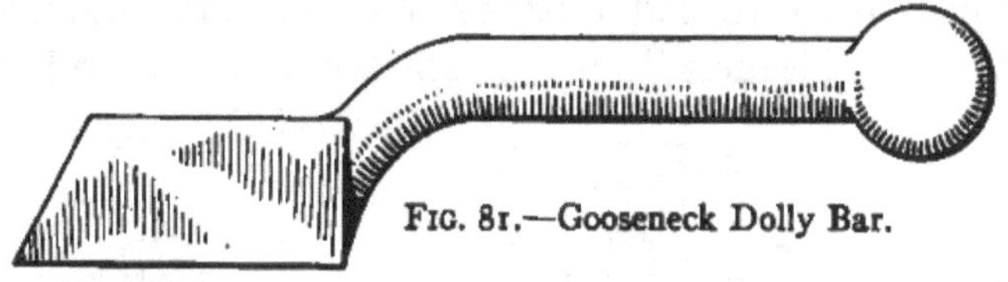
FIG. 81.—Gooseneck Dolly Bar.

places where there is little space behind the rivet. The tool weighs about 30 lb. and is used when the space is small and the air holder-on can not be fitted.

Straight Dolly Bar (Fig. 82): This is similar to the one shown above but is used when it is possible to get in

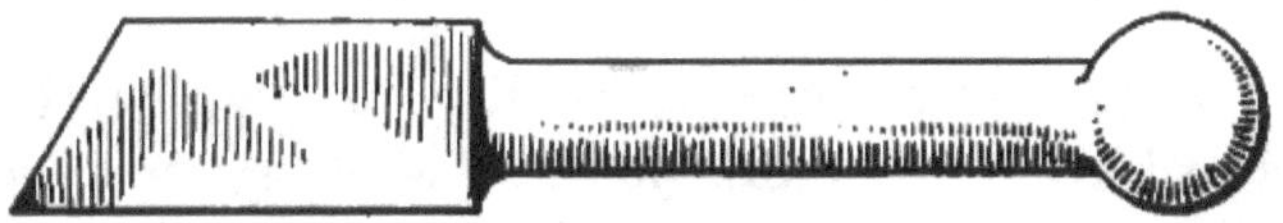
FIG. 82.—Straight Dolly Bar.

line with the rivet yet not convenient to use the air tool. (Both of these tools have different types of heads for the different kinds of rivets; as flush, button, and pan head.)

Rivet Forge (Fig. 83): This is used for heating rivets. It is a round pot of thick, cast iron, mounted on legs with a connection for an air hose. There is a portable tray above the air blast which forms the bed for the coke. Air passes up through the fire by means of numerous small holes in the tray. When the Heater Boy wishes to clean his fire, he turns on the air-cock wide open and the draft shoots the small cinders up out of the pot.

The sketch shows two handles which are used when hoisting the pot around on the ship. When the pot is to be cleaned, it is turned over on its side and dumped on deck. ("Cleaners" sweep up the cinders and carry them off the ship.)

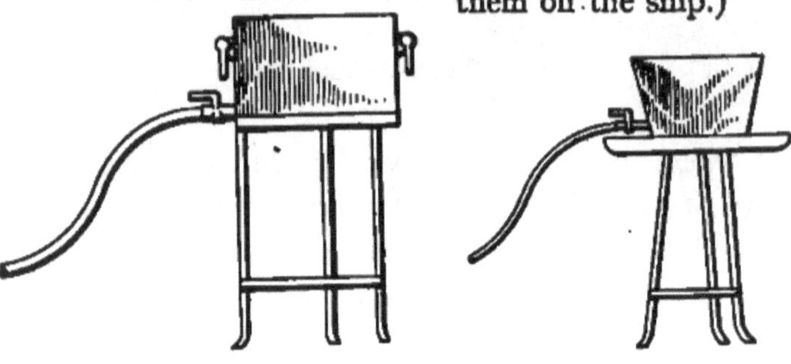

Fig. 83.—Rivet Forge.

Fig. 84.—Rivet Forge.

Fig. 84. This forge is similar to Fig. 83, but is made with a smaller fire pot and is of lighter construction. It is more easily handled and does not carry so large a fire.

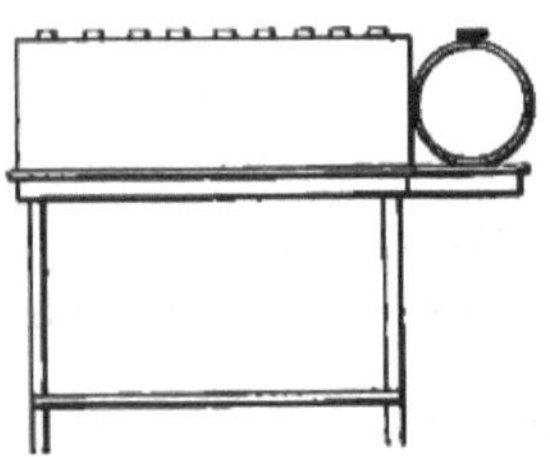

Fig. 85.—Rivet Forge.

Fig. 85. This oil rivet forge is shown as a sample of some now in use. The advantage which this type has over the coke-burning forge is in the handling of the fuel and the constant heat which is so easily controlled.

Rivet Testing Hammer (Fig. 86): This hammer is used in testing rivets to ascertain if they have been driven so that the material of the rivet completely fills the hole. It is customary to lightly hit the head of the rivet with the hammer at the same time that a finger of the other hand is placed on the side of the rivet head and also against the plate, thus enabling the tester to detect any motion between the two.

Fig. 86.—Rivet Testing Hammer.

Red Lead Gun (Fig. 87): This is a barrel with a screw plunger at one end and a $\frac{3}{8}$-in. pipe, with a nut, at the other end. When a leak is found and it is impossible to caulk it, due to the location, a $\frac{3}{8}$-in. tap is drilled and threaded, the "gun" filled with red lead putty. The gun is screwed into the tapped hole, plunger then screwed into the gun, thus forcing the putty out and into the space between the plates, or plate and angle, where the leak occurs. The $\frac{3}{8}$-in. hole is then filled with a metal plug, the putty hardens and stops up the leak.

Fig. 87.—Red Lead Gun.

Plunger (Fig. 88): The plunger of the Red Lead Gun has one end threaded to suit the inside of the barrel of the gun and the other end is square to fit the hand wrench, used when turning the plunger.

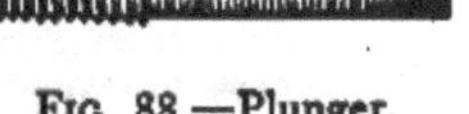

Fig. 88.—Plunger.

Hot Cutter (Fig. 89): This is used for cutting hot metal by means of a heavy maul, when doing hand riveting, or other work.

Fig. 89.—Hot Cutter.

Side Cutter (Fig. 90): This tool has a bevel edge and is used for side cutting with a hand hammer.

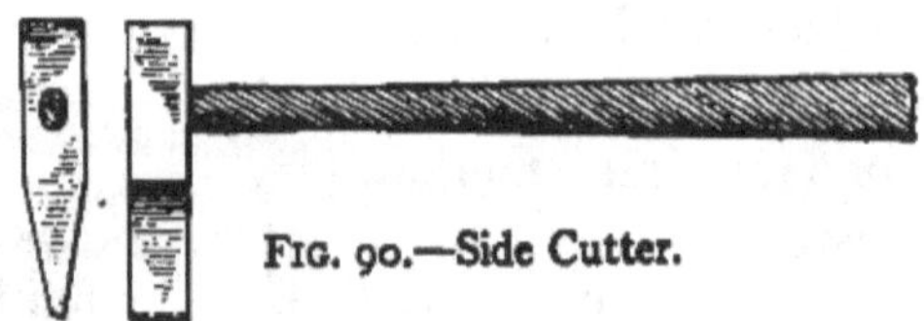
Fig. 90.—Side Cutter.

Cape Chisel (Fig. 91): This tool is used in an air hammer by chippers and caulkers for chipping work.

Fig. 91.—Cape Chisel.

Side Cutter (Fig. 92): This tool is used in an air hammer for chipping.

Fig. 92.—Side Cutter.

Straight Caulking Chisel (Fig. 93): This tool is used for ordinary caulking work where it is easily reached, and all straight work.

Fig. 93.—Straight Caulking Chisel.

Bent Caulking Chisel (Fig. 94): This is used for caulking in places which are difficult to reach with the straight chisel.

Fig. 94.—Bent Caulking Chisel.

Fine Caulking Chisel (Fig. 95): This tool is used for finishing and is also used for light work.

Fig. 95.—Fine Caulking Chisel.

Roughing Chisel (Fig. 96): This is similar to Fig. 93, except the face is knurled.

Fig. 96.—Roughing Chisel.

Roughing Chisel (Fig. 97): This bent chisel is similar to Fig. 94, except the face is knurled.

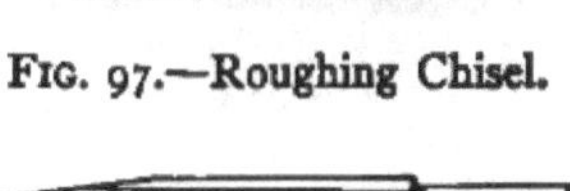

FIG. 97.—Roughing Chisel.

Roughing Chisel (Fig. 98): This chisel completes the set of Roughing-in tools which are used for preliminary work when the edge of the plate or angle is in poor condition for caulking. This tool (Fig. 98), has a rounded end, knurled, and is used before tool No. 95.

FIG. 98.—Roughing Chisel.

Straight Fuller (Fig. 99): This machine tool is used for finishing up a caulked joint on straight work.

FIG. 99.—Straight Fuller.

Bent Fuller (Fig. 100): Similar to Fig. 99, but is used for finishing up caulking when the air-gun must be held at an angle.

FIG. 100.—Bent Fuller.

Ripper (Fig. 101): This is used for opening up a seam or any straight line cutting in a steel plate.

FIG. 101.—Ripper.

Gouger (Fig. 102): This tool is used for destroying a rivet point, before backing out the rivet, or other similar work.

FIG. 102.—Gouger.

Bobbing Tools (Fig. 103): This machine tool is used for smoothing out surface work when caulking. It is a straight tool with an extra piece of rub-

FIG. 103.—Bobbing Tool.

ber hose (usually added by the operator) for a handle.

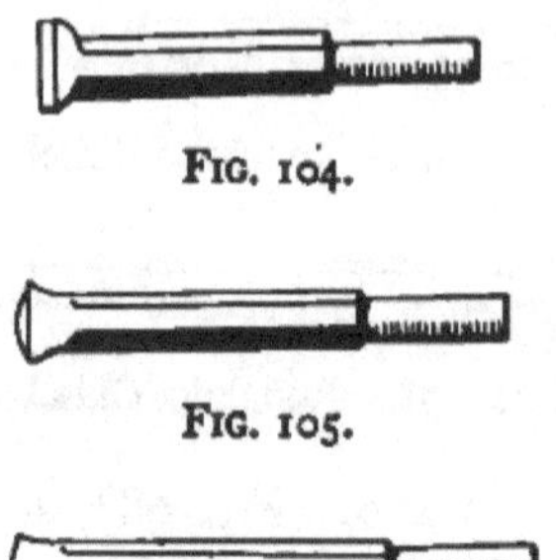

Fig. 104.

Fig. 105.

Fig. 106.
Bobbing Tools.

Fig. 104. Used by caulkers for caulking straight work.

Fig. 105. This tool is used similarly to Fig. 104 on rounded work.

Fig. 106. Another machine tool used for caulking, particularly on bent work.

Fig. 107.—Cold Chisel.

Cold Chisel (Fig. 107): This is a hand tool used for any general work required, as it is like the ordinary cold chisel.

Fig. 108.
Center Punch.

Center Punch (Fig. 108): This is the ordinary type used for marking the center of a hole to be drilled or punched.

Straight Caulking Chisel (Fig. 109): This tool is used for hand work on straight caulking. (Straight means where the work on the seam is all clear and easy to do, as regards any interference from other parts of the ship structure.) Similar to Fig. 93.

Fig. 109.
Straight Caulking Chisel.

Bent Caulking Chisel (Fig. 110): This hand tool is used for odd corners, around angle bars, under foundations, etc., where the straight hand tool would not work.

Fig. 110.
Bent Caulking Chisel.

Fine Caulking Chisel (Fig. 111): Used for finishing or light work. (These hand caulking tools are often used for testing tanks when the tank is full of water and leaks appear at different places. The hand caulking workman then travels over all the small leaks, and is able to caulk the metal by a few blows at a time, having more control over the caulking iron than when it is operated by air.)

Fig. 111.
Fine Caulking Chisel.

Diamond Point Chipping Tool (Fig. 112): This hand tool is ground down to a diamond point and is used for chipping a groove.

Fig. 112.—Diamond Point Chipping Tool.

Gouge (Fig. 113): This hand gouge is used for groove work giving a flat surface and is often used for a deep cut.

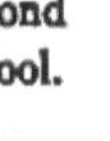

Fig. 113.—Gouge.

Cape Chisel (Fig. 114): Used like Fig. 97 for a finer cut.

Fig. 114.—Cape Chisel.

Fig. 115.
Straight Fuller.

Fig. 116.
Bent Fuller.

Straight Fuller (Fig. 115): This hand tool is used to finish up a caulked seam.

Bent Fuller (Fig. 116). This is used for finish work when the seam is in a location which is difficult to reach with a straight fuller. (Both of these Fullers are used for light work when a mere tapping is all that is necessary. They are more easily handled and more sensitive than the air tools.)

Hand Fuller (Fig. 117A): This tool is used with a heavy maul for creasing plates where a sharp knuckle is required. It is often used after the plate is in place on the ship. (Sometimes an acetylene torch is used to heat the plate locally, just where the knuckle is to be made.)

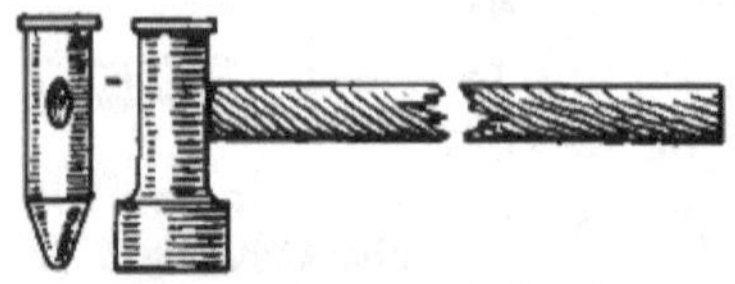

Fig. 117A.—Hand Fuller.

Hand Flatter (Fig. 117B): This is used to reverse the action of a Fuller. When a plate is accidentally creased or a correction is to be made, the Flatter is used to straighten out the plate and make it flat. A heavy maul and sometimes an application of heat is used, the same as when using a Fuller. (Both the Fuller and the Flatter will form the plate and make a finished job, whereas, if the maul itself were used, to come in contact with the plate, the plate would soon be scarred and present so poor an appearance that an inspector would not pass it.)

Fig. 117B.—Hand Flatter.

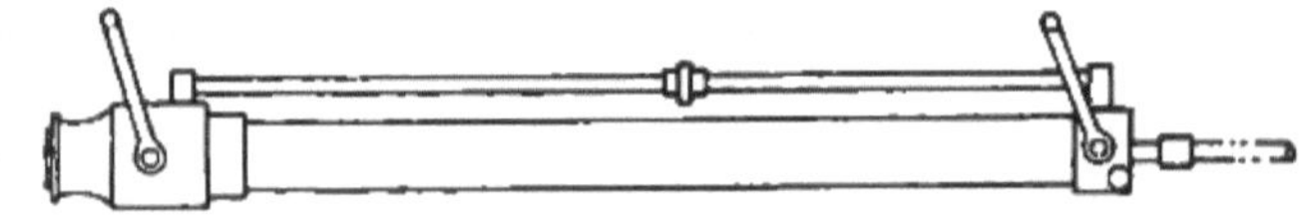

Fig. 118.

Rivet Buster (Fig. 118): A pneumatic hammer for driving the heads off rivets that have been condemned and it is desired to remove quickly from out the plate. It consists of a heavy air chamber and plunger, and works in a manner similar to the air hammer for driving rivets only it is in all respects much more powerful. This hammer requires four or five husky men to operate but when accustomed to working together they will accomplish considerable work in a day with the machine.

Side Cutter (Fig. 119): Used in the rivet busting gun when chipping off high head rivets.

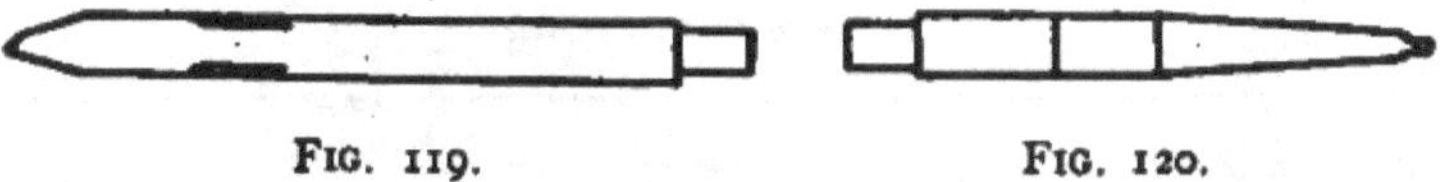

Fig. 119. Fig. 120.

Punch (Fig. 120): In the rivet busting gun this punch is used for backing out bad rivets.

Rivet Buster (Fig. 121): With an heavy maul this tool

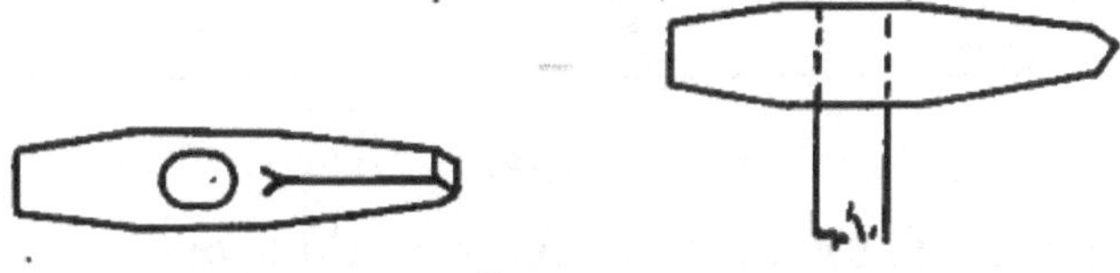

Fig. 121.

will remove bad rivets. This hand rivet buster is used with an heavy maul where the pneumatic buster cannot be used, due to small space.

Baldface Hand Riveting Hammer (Fig. 122): Used by hand workers for driving heavy rivets.

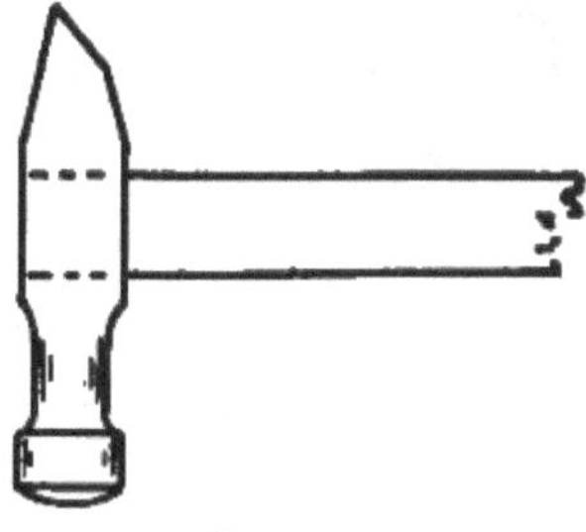

Fig. 122.

Bottom Rig (Fig. 123): For driving shell rivets, "upper cuts" on the bottom of the ship. This consists of an air hammer supported on the end of a pipe, held in position by means of two wheels which allow it to slide back and forth. One of the wheels has a standard which is held to the work by bolt-

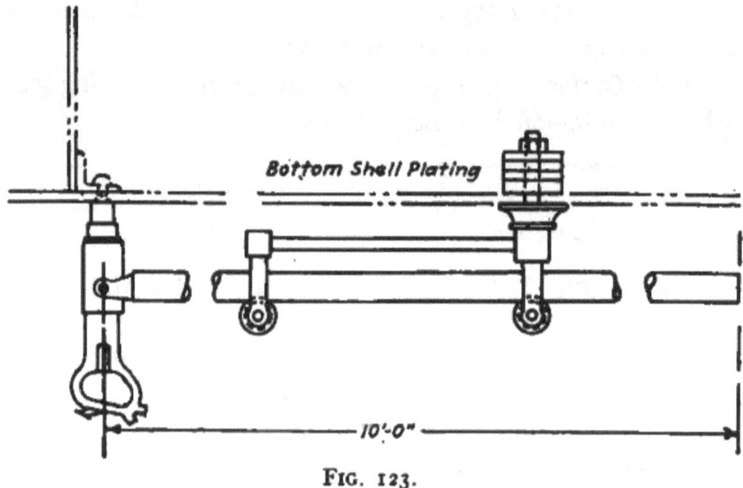

FIG. 123.

ing up through the plating. As it is held in only one place, it is possible to swing the hammer sideways as needed.

By using this rig the work is easy for the riveter as he only needs to guide the hammer and all strain of holding it

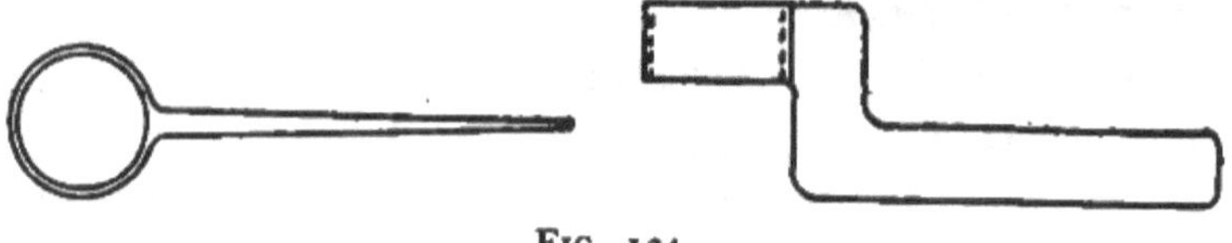

FIG. 124.

is taken up by the pipe. The bolt up through the shell plating is sufficiently long to allow for any ordinary thickness. Packing is used under the nut when necessary.

Ring (Fig. 124): This is a ring with a handle and is

sometimes used by riveters when working on the bottom shell. The ring fits around the end of the gun and it is possible to hold the gun in position against the rivet point without having the hand so close to the hot rivet as necessary when this arrangement is not used.

Notes on the Operation of Air Machines

Nearly all of the drilling motors used on the shipways are operated by compressed air coming in through the valve contained in the handle. By a simple wrist motion, turning the outside of this valve, it is possible to allow the compressed air to rush into the body of the machine, through ports which distribute it as desired. As the motion of the tool is rotary, it is necessary to have a crank turning the stem which holds the chuck for the drilling or reaming tool.

On some machines there are two pistons working in a dash-pot cylinder having one open end, and connecting rods from these pistons carried out to the crank. As the cranks are set at an angle, it is possible to avoid a "dead center." The lower end of the crank is fitted with a geared wheel so that the power is transmitted through a series of these wheels to the stem holding the chuck. When using this motor in drilling, the feed screw handle, which is shown at the top, is operated to carry the end of the screw up against the backing which is needed to take the thrust of the drill. (Fig. 125.)

For ordinary work on ship-board such as drilling new holes or reaming punched holes, the "round machine" is generally used. Fig. 126 show such a machine with the

FIG. 125.

FIG. 126.

valve handle and "dead handle." The valve handle, the shorter of the two, is operated by the man in charge and to this the air hose is attached.

A "corner machine" used for drilling and reaming when in a corner is illustrated in Fig. 127. The difference in the shape of this from the round machine is due to the fact

Fig. 127.

that the chuck for the drill is at the extreme end, this being necessary to make it possible for the workman to get the point of his drill into a small space. It is operated in the same way as the other motor but has no dead handle.

Fig. 128 shows a typical "chipping hammer." The design of these varies with the different manufacturers but the main principle is the same in all cases, both the riveting and chipping hammers performing similar functions, although the design must be altered somewhat due to the different uses of the tools.

The chipping hammer gives a lighter blow than the one for riveting and has shorter stroke of the plunger. The forward end of the chipping hammer holds the chisel firmly

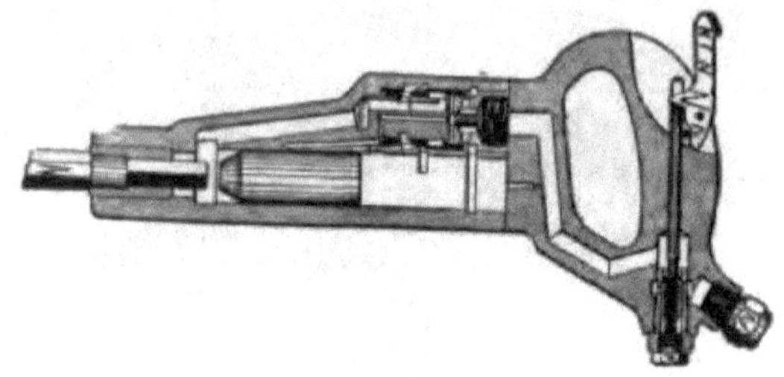

FIG. 128.

in position and allows plenty of shank to reach in to receive a full blow from the plunger.

The forward end of the riveting hammer is designed to hold the shank of the die and the plunger hits directly on the end of it. These dies are generally made with a square end and are ground according to the custom in the ship yard where used, the "concave" die being hollowed out with a sharp edge or rim. (It is customary to use these where the rivets are driven finished, and caulked against the plate, at the same time.)

The "flush" die has an end square with the body of the die.

The "baldface" die has an end which has been slightly rounded.

These dies are all used on flush riveting according to

the practice of the individual workmen. In both of these hammers the air passes in at the under part of the handle and is allowed access to the interior chambers by means of a valve in the handle. This valve is operated by the thumb of the hand which grasps the handle. The downward pressure on the trigger drops the valve, against the pressure of the spring, and the compressed air rushes in through the handle into the body of the hammer. (Fig. 129.)

In those designs where the lower part of the handle is open, this passage is carried around through the upper part.

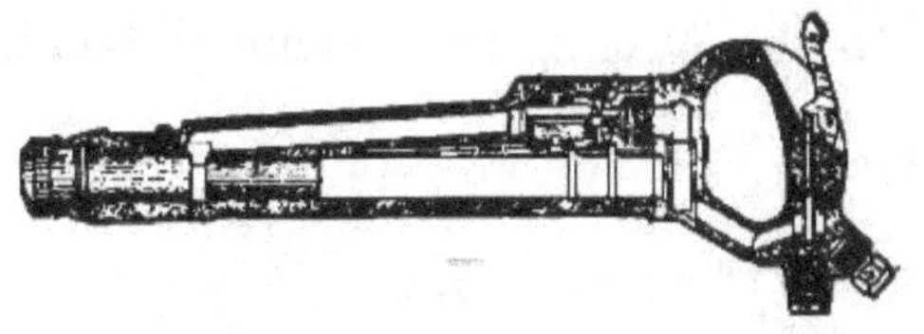

Fig. 129.

From here the air goes directly through the valve, which is often made up of the valve and valve guide or stem along which it travels.

In some makes of hammer the strainer is located in this valve and stops the passage of any dirt or grime, thus preventing any damage to the working parts of the tool; the strainer and valve can easily be removed for cleaning.

From the valve the air is allowed to pass into the chamber containing the plunger, and pushing on the end of the plunger forces it either toward the rear of the chamber on a back stroke or toward the front end, hitting the die against the rivet point according to the location of the plunger at the time the air is admitted. The passage of the plunger acts as a slide valve in an engine and the air is alternately cut off from one end and allowed to come into the other end so that the plunger is forced to make rapid strokes in both directions.

After the air has done its work on the plunger it exhausts out through a port in the side of the gun. The size of these guns vary according to the class of work to be done (size and point of the rivet) and the force of the blow to be delivered.

The Grinders are operated by compressed air in a manner similar to the Drilling and Reaming Machines. Air is

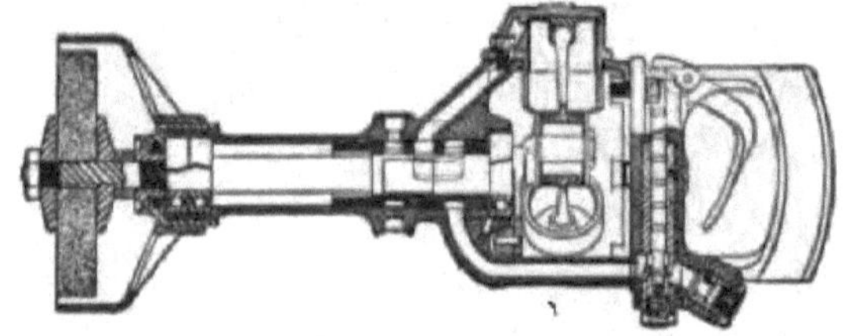

FIG. 130.

admitted by means of the valve operated by the trigger and as it rushes into the dash-pot cylinders sets them in motion and causes the spindle holding the grinding wheel to turn at a high rate of speed. The exhaust port for the air is located conveniently on the side so that after it has done its work in the cylinder it is allowed to escape into the atmosphere. (Fig. 130.)*

*The sectional views in Figs. 125 to 130 inclusive are shown through the courtesy of Ingersoll-Rand Company.

CHAPTER IV

Shipway

The location of a ship yard and the layout of its Shipway are determined in a great measure by the breadth of the stream on which it is situated and by the value of the land on it.

Most of the shipyards have sufficient breadth of water surface so as to be able to launch their vessels end-on (stern first) but some shipyards being located on narrow rivers or creeks are not so fortunate in the space needed and must lay down their work so the vessel may be launched sideways.

When ships are built in a row with one end toward the water, more of them can be built at one time than if they were built broadside to the water front. To illustrate, if the shipways are 100 feet between centers, then five ships could be built at the same time instead of one ship with a length of about 500 feet which was built parallel to the water's edge and was to be launched sideways.

For that reason it is customary wherever possible to build the shipways at right angles with the shoreline.

As the majority of the shipyards are bought and laid out with these conditions in view, it follows that in most yards the shipways are laid out with the vessel's end on to the water and the following description is for that type of yard.

The Shipway is the platform on which the hull of the vessel is laid down and from which the completed ship is finally launched into the water.

The shipways generally consist of a wood deck of planking carried on piling, where the ground is soft, for the whole length of the ship, or where the ground is firm, for the inboard end, under the bow.

The foundation is laid out in rows of piles or "bents" extending across the shipway and surmounted at the top by a heavy head "stringer" which is carried level. These bents vary in height according to the inclination of the ways and are spaced at a constant distance (Fig. 131) according to the weight of the ship as heavier ships require a distance between bents to be less than in vessels of smaller dimension.

The main framing, longitudinally, consists of two heavy stringers approximately 3 ft. apart, equidistant from the center line of the shipway and extending the full length. The deck used for walking space is carried

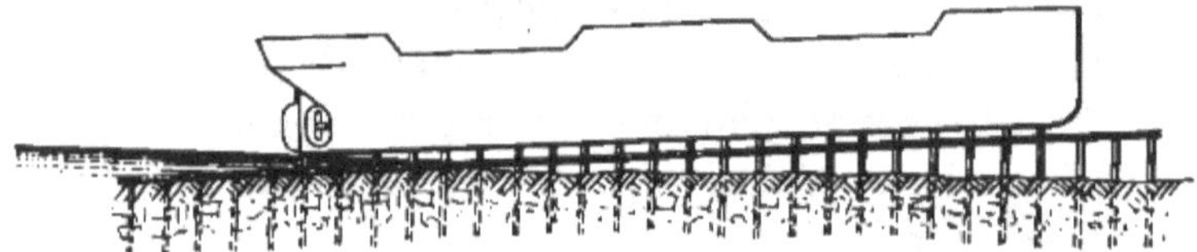

Fig. 131.

in a line with these stringers and is spiked securely to the top of the cross bents.

The outboard or water end of the shipway is carried down to the ground about high-water mark in order to have the extreme outer end below the high-water mark and to reduce the height of the inboard or bow end where the piling has been put in for the foundation.

On these two main stringers near the center line are laid the keel blocks. These keel blocks are built in size according to the weight of the vessel to be carried. (Fig. 132.)

These blocks are built up to give the required height between the bottom of the ship and the deck of the shipway. They should allow a man to stand underneath the bottom of the ship and use the air-riveting hammer and do the other work of caulking, chipping, etc., without having to stoop too much.

FIG. 132.

As all vessels must be built in a position from which they can be easily launched into the water, it is necessary to have them inclined so that they can be readily moved by sliding.

If they were to be built level, force would have to be exerted to move them into the water, but with the inclination provided, they will move of their own accord.

The inclination given to the shipway varies with the weight of the ship to be built and launched; it is from $\frac{1}{2}$ to $\frac{3}{4}$ in. per foot.

In some shipyards where the vessels are launched into a river having a swift current, it is necessary to give considerable inclination to the keel blocks so that the speed of the ship when launched will be sufficiently fast to enable her to move quickly into the stream before the current of water can twist the boat on the shipway.

In most of the shipyards, this is not a necessity, as the launching is different in tidal water. The time of launching is arranged to take place at exactly high tide when there is the greatest depth of water and it is stationary.

The shipway just described is for the modern shipyard where fast work is one of the necessities; yet many of the wooden vessels that ply along our sea coast were built in the owner's back yard. This is especially true in those "shipyards" along the shores of the Kennebec and other rivers in Maine and other States, where the wooden vessels are framed, planked and launched by men who have only their hand tools, as saws, broad axe, adze, etc.

In these "yards" the shipbuilder lays out his keel blocks to the required inclination and is often favored by the sloping banks of the river. In this way it is possible to secure the correct position of the keel without building up the keel blocks under the bow as is necessary when the ground is more nearly level.

By this arrangement it is possible for the workmen to remain standing on the ground and yet be able to reach their work on the under side of the bottom of the vessel.

Under the heading of "Launching," the necessity of the wedge blocks will be clearly shown. The cap block on top is beveled to an inclination which gives a line parallel to the inclination of the shipway. (The keel blocks stand vertically.)

After the ship carpenters have given the required

beveling according to the sight laid out by the shipyard engineering department, the engineers also furnish the carpenters with a sight for a center line which is struck down the middle of the line of cap blocks.

This sight line is scrieved out into the top surface of the blocks, across the edge, and a little down the sides, in order that the exact point of center line can be determined when standing below and looking up at the bottom of the keel plates.

Spur shores are set against the back side of the cap block down to the forward side of the next bottom block. These are to prevent sliding motion of the blocks downhill as the keel is being laid.

The mid-ship point of the ship is laid on the shipway by the engineers and the " dead-flat " mark is established on the stringer by a cross mark.

Sometimes the ship carpenters will set up joists as a " monument " to give the exact location for the first keel plate, having a monument at one end and two or three on one side; thus locating exactly the right spot in a fore and aft line and the side monuments giving the exact center of the first keel plate as marked on the line of the keel blocks.

Some shipyards use "pin shores" for supporting the bottom plating and in this case many of these are cut from the upper end of piles, stripped of the bark, measured for the right length and laid away until ready for use. To get the proper height, double wedges are used at the lower end of the shore and these are also made and stored away until ready for the laying of the keel.

Other shipyards have "spalls" consisting of a heavy plank carried out on each side of the keel blocks and supported at the outer end by an upright. These spalls are fastened to the keel blocks and extend the full breadth of the vessel. Both systems are illustrated in Fig. 133.

For rapid work, this system of using spalls is preferred as it is possible to lay considerable of the bottom of the shell plating nearly in the place where it is to be finally located and can be done before the actual work of laying

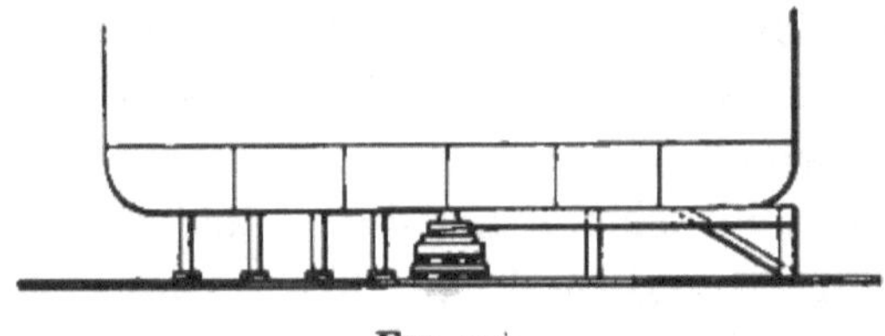

Fig. 133.

the keel is commenced. These spalls are later replaced by shores, which can carry more weight than the spalls.

Where the system of pin shores is in vogue, the bottom plating cannot be laid until after the keel has been placed on the keel blocks, thus delaying the work and sometimes making it rather hazardous because, unless the tops of the pin shores are secured by a heavy nail or spike through a rivet hole they are apt to become dislodged by a jar and fall down on the shipway thus leaving that part of the bottom of the ship unsupported.

The method of handling material while building the ship varies in yards according to the size of work, amount of capital first involved when building the yard, and other considerations. Some shipyards have a system of overhead traveling crane running the whole length of the shipway and extending in line sufficiently far so that any bulkheads which have been built on the ground can be raised and carried over the boat and lowered directly into place. Another method employed has a line of steel towers on both sides of the shipway which carry a "stiff-leg" derrick; in this case, there would be three or four of these towers on each side of the ship varying with the layout of the vessel. Other yards have different systems according to the layout of the ways and vary more or less in a modification of these two types.

Some of the more recent yards have covered in their shipways by a light steel roof; thus providing the workers with a chance to continue in stormy weather.

Some of the western shipyards are fitted with cableways which consist of a series of wire ropes and run parallel directly over the two sides of the ship, each pair of cableways being separated at the two ends of the boat by tall masts well supported by back stays. An electric motor operates the traveling block which can carry and deposit the steel plates as desired. When the material is to be laid they are sometimes both used at the same time, each carrying a part of the load; in this way, all parts of the ship can be reached and it facilitates the rapid handling of the material.

The air hammers and drills are supplied by an air main of steel piping which is carried along on each side of the shipway with branches running up the side of the boat and leads going in onto the deck. These air mains are supplied with manifolds on the shipway; usually about four or six on each side of the center line for use on the bottom plating. Manifolds are fitted where the branches lead in onto the decks and these are fitted for a number of hose couplings. Each manifold has shut-off valves. . (The rubber hose leading from this manifold to the riveting hammers and other air tools has been taken up under " Shipyard Tools.")

In order to make it convenient for work on the outside of the ship, it is necessary to surround the boat with "staging." This consists of two uprights (built up of square timber), the space between has a spall with one end hanging out like a cantilever beam. On the outer end of this spall, two staging planks are laid, upon which the men can walk or stand at their work. This staging is placed about 16 ft. apart along the ship so that the planks can run from one stage bent to the other and can have a little additional length for overlapping.

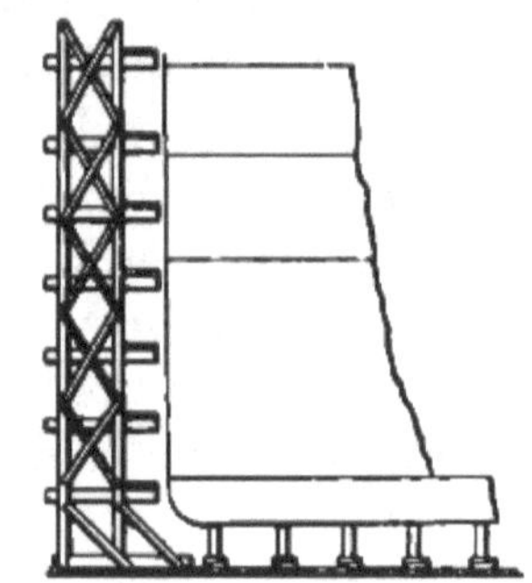

Fig. 134.

The staging generally is built in two or three sections, according to the height, and the spalls are placed about 6 ft. apart (vertically) so that a man can easily work on one plank without interfering with the one over his head. In this way, it is possible for the men to reach any part of the side of the shell plating as these spalls extend close to the ship's side. The spalls are portable and can be raised or lowered as desired. (Fig. 134.)

The staging on both sides and around the bow can remain in position after the boat has been completed and is ready for launching but the staging around the stern must be removed in order to allow the vessel to slide down the shipway and into the water.

Access to the ship is gained either by stairways built in with the staging or else by means of a "runway." The stairways are regular stairs going as high as possible and then stopping in a landing at one of the rows of stage planking, along which the men walk until they come to the next flight of stairs.

The runway is generally made up of planks laid close together forming a platform about five planks wide and inclined as steep as possible, running from the ground, near the forward end of the ship and reaching the height of the top deck about mid-length of the ship. Cleats are nailed on at the proper distance for an easy step. A hand rail prevents workmen from being crowded off while passing each other, some going up as others pass down to the ground.

CHAPTER V

Keels

When sailing ships were first designed it was found necessary to build a projection along their length, under the bottom, to prevent the ship from "falling off" the wind, i.e., sliding sideways due to wind pressure.

This projection was made by extending the depth of the keel for some distance below the bottom of the ship, thus forming a wall which would catch on the water when moving sideways, and as it was narrow as well as deep it did not obstruct any motion in a forward direction.

Wooden ships of to-day are built with such a keel, a thin wood "shoe" or wearing strip on the bottom edge, and on the top edge inside the ship securely fastened to the keel proper is another similar member as the keel called the "center line keelson." The keelson is for additional strength. The "garboard" strake of wood planking is let into a "rabbett" in the side of the keel.

When the first iron steamers were designed the same idea of a keel was developed in the form of a "bar keel." This was a heavy forging, deep and narrow, and the garboard strakes of the bottom plating of the shell were flanged down against the side of the keel and then "through riveted."

As the naval architects advanced in the building of

iron and then steel vessels this bar keel was replaced by another design, due to the large size necessary in big vessels and also due to the fact that if the ship did happen to go aground this projection was liable to cause much trouble as it would take too much weight of the ship, locally, and start leaky joints. Then again it had been found that vessels under steam power acted differently than those under sail and that the projection below the hull in the form of a center line keel was not necessary.

Acting on this knowledge the marine designers used a heavy, flat (longitudinal) plate keel and carried an inner, vertical plate on the center line of the ship to maintain the necessary stiffness and strength. This vertical plate is called the center line keelson.

Thus the outside of the bottom of the ship has no projection and if the ship takes ground the weight of the vessel is sustained by a large surface with less danger of straining.

Some designers prefer to use one heavy plate for the flat keel while others use two, thinner, lighter weight plates. There are arguments in favor of both designs, but it is sufficient to mention here only the fact of the two types.

The necessary riveting to secure the one heavy keel plate to the ship's framing is used also to hold the two plates together when the keel is formed in that manner.

The lower or outer p'ate is called the outer keel, the upper or inner one is called the inner keel. The outer keel is sufficiently wider than the inner keel to allow for a lap of the shell-plating.

The outer keel extends for the full length of the ship

and is finished at the forward end by a "dished" or "furnaced" plate which connects directly to the stem casting.

The after end is connected directly to the stern frame and varies in shape according to the design of the stern.

The inner keel generally extends about two-thirds of the ship's length, where the greatest strength is required, and is placed in the ship as a strength member.

The keel framing is completed by adding a vertical plate, the width of which determines the depth of the inner bottom. This vertical plate is called the "keelson." These plates are the backbone of the ship structure and it is necessary when laying down the ship to have these members correct in alignment and measurement, fore and aft, as the correctness of the succeeding erection of material will depend upon the care with which the start has been made. (Fig. 135 and Fig. 136.)

When laying the keel, the start is made with the midship plate of the outer keel, this being placed over the "dead-flat" mark and against the monument as described in the chapter on "Shipway."

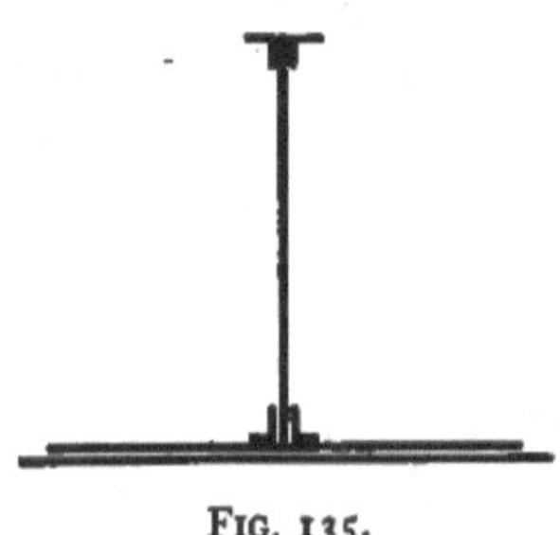

FIG. 135.

The bottom side of the plate is usually lined off with a chalk line giving the exact center line fore and aft. This line must coincide with the center line as scrieved in on the keel blocks.

The laying of this plate is followed by succeeding plates

working both toward the stern and toward the stem at the same time, each plate being carefully laid to keep an alignment of all plates perfect. In order to retain the outer keel plates in position until the inner keel plates are laid, it is well to bolt on a small, flat plate (one bolt) at each corner of alternate plates to hold those plates in between from sagging. To hold the plates together (lengthwise) a small angle clip is bolted (one bolt) back of the flat clip just mentioned, and a bolt about 12 in. long, with nut, is carried to a similar clip on the next

Fig. 136.

plate. All of these clips are placed on the outer keel at the landing for the garboard strake so they will not interfere with laying the inner keel.

After all the outer keel is down, a steel tape should be used to check the fore and aft distance to be sure that the frame space will come right as sometimes in fabricating the plate, it may happen to be a little too long or too short and a small fraction of an inch at this place in the work will cause considerable trouble with shell- and deck-plating later on in the construction.

After the outer keel has been laid and "regulated"

the inner keel is laid on it and if correctly laid out in the mould-loft, all rivet holes should come directly in line.

After these plates have been laid together, they are bolted; thus holding them securely as one piece of metal. The keelson is now ready for placing and is laid with the rivet holes of the bottom angle bars matching those in the keel plates, the location of the keelson plates being taken from the plan as sent out from the drafting room.

In many of the yards, this keelson is laid after having been partly fabricated by having the holes for floor clips punched or else the clips riveted and the keel bars in place.

It is customary to have the keelson water-tight throughout nearly its entire length and the caulking is generally done on the starboard side.

In order to insure perfect water-tightness, it is customary to use " stop-waters " at certain places where there is liable to be a chance of seepage of water due to poor workmanship or where it is difficult to caulk as described in the chapter on "Stop Waters."

The keelson is now shored up temporarily at the top and the shipbuilders are ready to begin laying the bottom of the shell plating. The keelson plate is about 4 ft. high, according to the designed depth of the inner bottom. It is usually lap butted, treble riveted and has two angles at the upper edge and two angles at the lower edge. These angles are fitted in long lengths and have " splice bars " or " bosom " pieces at their butt joints.

CHAPTER VI

Shell Plating

A ship derives its main structural strength from its hull plating, which is divided into two main classes, bottom plating and side plating. The bottom plating extends across the bottom of the ship on both sides of the keel, while the side plating is from the curve of the " bilge " up the ship's side. Plating is laid in " strakes " the first being nearest the keel. This is often called the " gar-

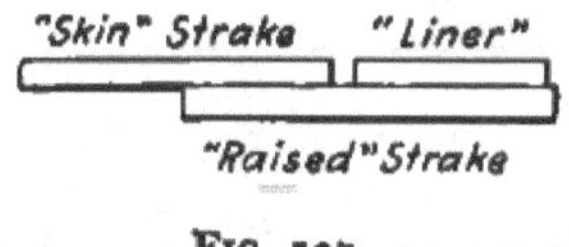

Fig. 137.

board strake." For convenience in working it is lettered *A*, the next strake being the *B* strake, the succeeding one being the *C* strake, and so on. Each plate is numbered, as: A5S, for *A* strake, plate number 5, Starboard Side: or C9P, meaning, *C* strake, plate number 9, Port Side.

Shell plating is laid either "raised" or" joggled." (On naval vessels and yachts it is often laid "flush," with the edges in line and an extra, narrow plate called a "buttstrap," carried behind the joint of the two plates. This strap is riveted to each plate, thus forming a continuous member.)

The "raised" plating has every other strake against the framing with the strakes between "raised" sufficiently to overlap onto the other strakes which are often called "inner" or "skin" strakes (see Fig. 137). Fig. 138 illustrates the use of "liners" which make possible the riveting of the raised strake to the frame without drawing the strake toward the framing and distorting its shape. The space between the raised strake and the framing is filled with a small distance piece called the "liner," which is the same width as the flange of the frame, the same thick-

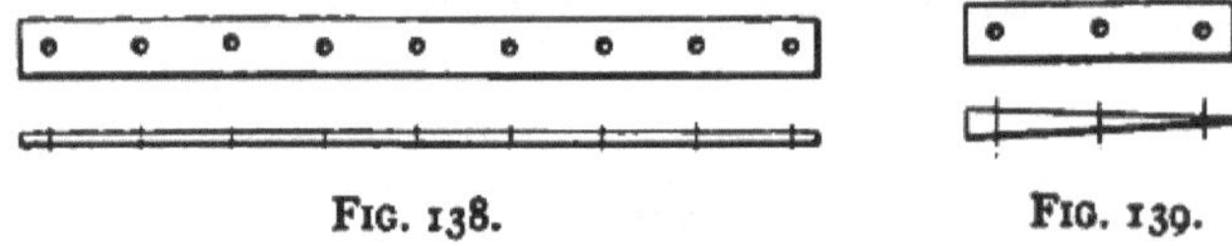

FIG. 138. FIG. 139.

ness as the skin plates and sufficiently long to run from one skin plate to the next.

"Tapered liners" are fitted on the landing or seam directly at the back of the butt-lap where one plate has been raised above the level in order to lap over the one which it joins at the end. The triangular space caused by the raising of the plate is filled by this type of liner. The butt end of the liner is the same thickness as the strake which it fits against and tapers to nothing at the other end; the width is governed by the width of the landing or seam of the plating and the length of the liner is determined by the number of rivet holes required in it by the inspector. Usually at least two are called for (see Fig. 139).

In order to avoid the use of tapered liners, in some shipyards the ends of plates along the longitudinal seams or landings are "scarphed." The method used for hand work in making the scarph is described under "Duties of the Chipper and Caulker" in Chapter II. A machine has been recently patented by a firm on the "West Coast" which forms these scarphs quickly and evenly.

The system of joggle plating differs from the one just described in that the raised strakes are offset when they overlap the skin strake and set directly against the framing of the ship, thus doing away with the need of a liner. (Fig. 140.) The joggle or "crimp" is put into the

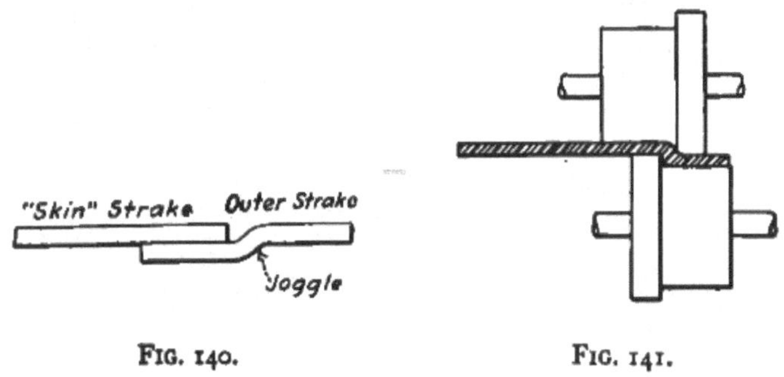

Fig. 140. Fig. 141.

plate by two wheels, one out of line with the other through which the plate is forced as indicated in Fig. 141. The amount of joggle is adjusted to suit the thickness of the plates to which it is to be fitted. The edge lap is called a "seam" or "landing" and is usually double riveted. The ends of the plates are butt-lapped; when this is done it is necessary to carry the forward end of each plate underneath (inside) the after end of the plate just ahead of it. This brings the after end of each shell plate on the outside of the ship and avoids unnecessary friction as the ship travels through the water. For the same reason all riveting in the shell below the deep load line is always "pan head" and "countersunk point." The button-point type of rivet being less expensive, is often used above the waterline, where smoothness is not so necessary.

The shell plating varies in thickness according to its location, being thickest amidship and tapering in thickness toward both ends of the ship. This preserves the uniformity of strength of the ship and avoids the tendency of its breaking apart in a heavy seaway, which is always greatest

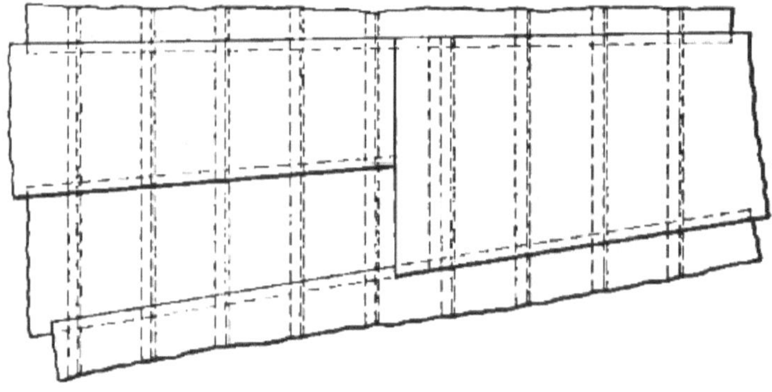

Fig. 142.

in the middle of the ship. Toward the forward end of the boat where the girth becomes less the width of the plates are contracted and usually two strakes are run into one as shown in Fig. 142. This plate is called the "stealer" and is used to prevent the plates from coming to a point. In some ships the plating is designed so that only one or two stealers are necessary and these are close to the ends of the ship and below the water line.

The **garboard** strake is next to the keel and the **bilge** strake is at the turn of the frames or the curve between the bottom and side plating. The **sheer** strake is the strake running next to the main deck of the ship. In some designs, these three strakes are given a little more thickness to increase the strength of the boat.

After the keels have been laid, the next step is to place the bottom plating starting with the *A* strake and working out toward the bilge of the ship.

Both sides of the shell plating are alike as far as location of butts is concerned, except on the *A* strake where the butts do not come in the same frame space. These butts are usually laid off to allow at least two frame spaces between butts on the *A* strake port and the *A* strake starboard. This is done to avoid a " brick shift " of butts, because of the structural strength.

This same arrangement of at least two frame spaces between butts is carried throughout the shell plating. It is necessary to have one intervening strake between butts of the two strakes each side of it. (Fig. 143.)

Places in the shell where locally an unusual number of rivets is required, are reinforced by means of a " doubling " plate to maintain the necessary amount of area in the material. This often happens at water-tight or other places where the rivets are close together, or when a hole is cut in the shell plate.

All butt laps are at least double riveted; the ends of the forward plates are laid on to the stem casting and secured by through rivets; the same being true at the aft end of the boat, according to the design of the stern post. This design is affected when a single, or two propellers are required.

After the bottom plating is in place, the floors are laid and side framing erected. The Erectors (or Plate Hangers) lay the side plating, putting on the skin strakes first and then the raised strakes. This side plating must

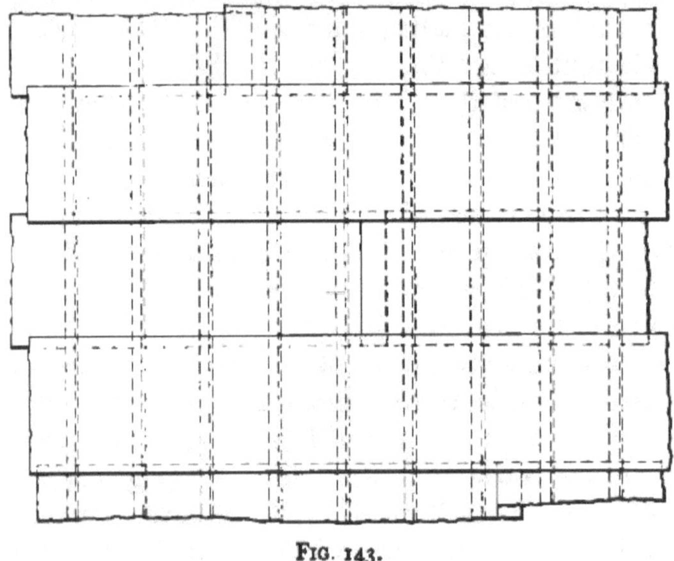

Fig. 143.

be securely bolted in position to prevent settling, because of the weight. Whenever such settling starts, it is very difficult to " regulate " the plating higher up as the work progresses; so each plate above the bilge must be securely fastened before the next is laid.

Some ships are fitted with a " bilge keel." This is fitted on the turn of the bilge of the ship and extends for about two-thirds of the length. Its duty is to prevent excessive rolling of the ship in a heavy sea as it catches against the water and retards the motion when the ship rolls. The usual type for merchant vessels is a Tee bar riveted to the shell plating and a medium weight of plate about 15 in. deep riveted to the standing flange. The

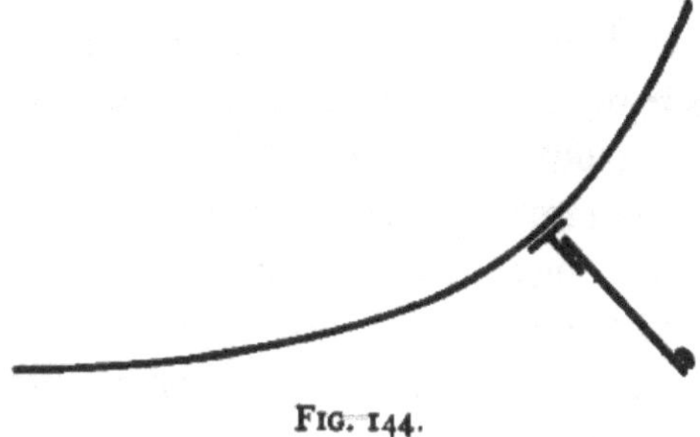

FIG. 144.

outer edge of this plate is reinforced by a half-round iron. (Fig. 144.)

The " oxter plate " is one of the most difficult plates in the shell plating to fit, due to its location. It is fitted just under the knuckle of the counter of the ship, where the upper part of the rudder post enters the shell plating. This plate is usually made in a short length for easy working.

The " boss plate " which is fitted around the boss where the propeller shaft emerges from the shell, is another plate which requires expert handling.

Where the forward end of the flat keel joins the lower end of the stem casting, a plate is fitted which is difficult to work, due to the abrupt change in form, as the forward end of this plate is carried around the stem and riveted to it, while the after end is flattened to take the curved surface of the next keel plate. The extreme curve gradually dying out in the length of a few plates to the flat shape of the remainder of the keel.

These plates just described which take an excessive change in form are called " dished " or " furnaced " plates, because they are worked by a blacksmith over a fire and bent to suit a templet.

(It is a good plan to leave off one outside plate near the bow and another near the stern, on the opposite side, for access to the Tank Top during the construction of the ship.

These plates are "hung" and regulated, then removed until it is nearly time for the ship to be launched, when they are put in place and riveted. Being outside plates they do not disturb the adjoining plates when they are finally secured.)

CHAPTER VII

Frames

In a ship the framing may be said to correspond to the ribs of the human body, as it resembles them both in utility and position. There are two types, " transverse " and " longitudinal."

The " transverse " system carries the main strength athwartship and prevents crushing in of the sides of the ship. It supplies the necessary strength to retain the form against the strains and stresses which occur when the ship is in a heavy seaway. The necessary longitudinal strength being supplied by the decks, shell, intercostal longitudinals and center line keelson.

The "longitudinal" or "Isherwood" system performs the same service in retaining the shape of the ship by means of girders which run lengthways of the ship. It is patented and coming into extensive use, but the transverse system is the one which is, however, most commonly used.

The "Isherwood" system is briefly described at the end of this chapter, which is devoted almost entirely to a detailed description of the "transverse" system because it is the one in general use in many of the shipyards.

Imagine a wooden model of the hull of a steamer, sawed into sections directly across the boat, the distance between these sections being quite close and equally spaced. When one of these sections has been removed from near the center of the ship, it will have a shape like Fig. 145. Suppose the section near the bow is removed, it will have a shape like Fig. 146; the section near the stern like Fig. 147.

The drafting room of the Shipbuilding Company prepares a table of "Mold Loft Offsets" giving dimensions

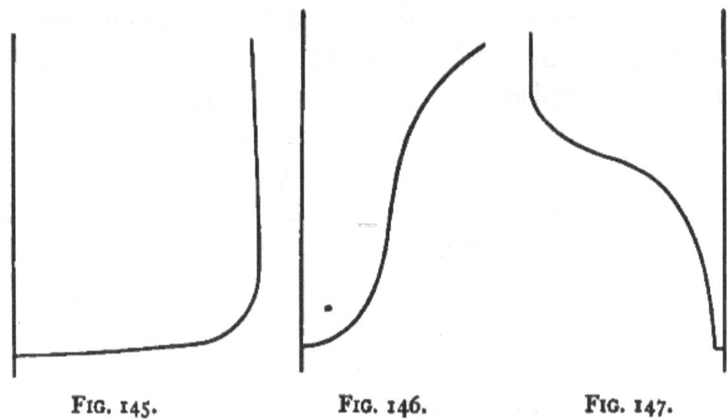

FIG. 145. FIG. 146. FIG. 147.

from which it is possible to reproduce transverse curves at any place in the hull of the ship. These curves are taken at points along the center line of the ship, which correspond to the location of the transverse frames. The Mold Loft receives this Offset Table and prepares a scrieve board for the body plan which represents all the frames drawn the full size. (Fig. 148.)

From this scrieve board, the template maker furnishes templates of thin wood (about one-eighth of an inch thick) which gives the outline of the curve of the frame. There is a separate template made for each frame which varies in shape. As much of the middle body of the ship is

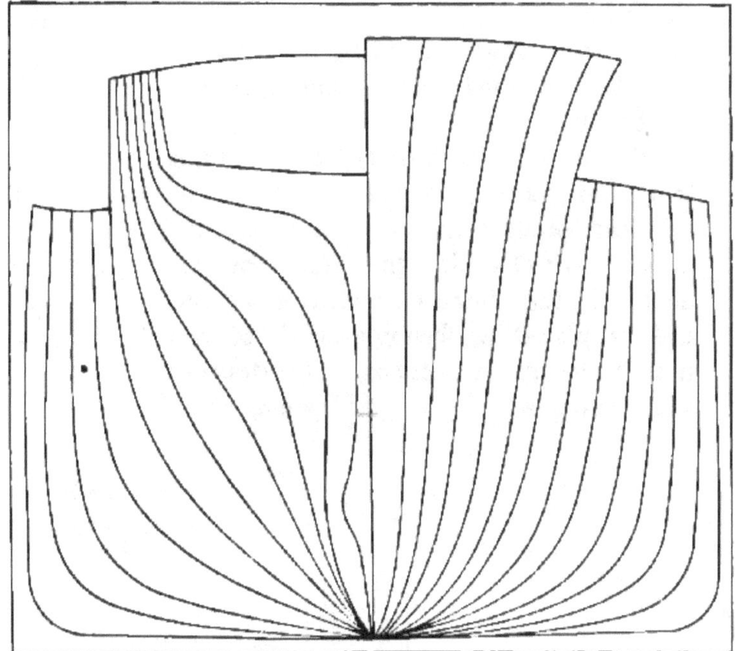

FIG. 148.

carried alike as possible in order to duplicate the frames in the "midship" or "dead-flat" part of the vessel.

From these wooden templates, an iron template (called a "set iron") is made of thin flat bar, duplicating the shape of the frame. It is necessary to use wood for the template made on the scrieve board. Because of the heat, the frame angle or channel can not be laid against the wood template (but a thin, flat bar can be laid on the wood without burning it), hence a steel template is made from the wooden one and the steel template is used to form the frame.

FIG. 149.

Fig. 149 shows a mold for a side frame with the landings of the shell plating marked on the edge.

The furnace for heating the frames is long and narrow, is heated by oil fuel and can produce a very high temperature. Directly in the front of the door of the furnace, which is at one end, is placed a "bending slab" of cast iron, made up of sections, each designed with "cross webs," giving an opening about

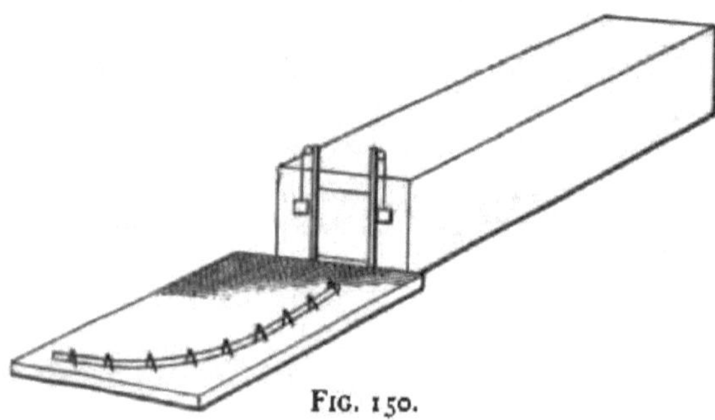

FIG. 150.

2 in. square. These sections are placed touching one another, forming a bending slab of sufficient size to hold an entire frame. (Fig. 150.)

The material to be bent to the form of the frame, whether channel bar or angle bar, is shoved into the oven and allowed to remain there until heated to a bright red color.

The steel template is secured to the bending slab by means of bent " dogs," one arm of which passes into the

Fig. 151.

slab and the end of the other arm grips the template on the top surface. After the template is securely fastened to the bending slab, the frame is drawn out by means of tongs, hauled close up to the template and the last end out from the furnace is secured against the template by means of a bent " dog." (Fig. 151.)

The hot frame is now bent around against the template by means of levers with a long handle having a pin on the lower side which will fit in any of the openings in the bending slab and the iron extending beyond the pin as a " cam shaped " face. (Fig. 152.)

By means of this " cam," the increasing distance from the pin forces the hot frame into place.

As the frame takes the form of the steel template, it is secured in position by other " dogs," which hold it until it is cooled in that shape. After the frame is sufficiently cooled to retain its shape, the "dogs" are knocked loose and the frame is taken out into the yard, clear of the bending slab. While they are working on frames amidship, all of a similar shape, the same steel template is retained in position. When the other frames at the ends of the ship are being bent, the templates must be changed each time.

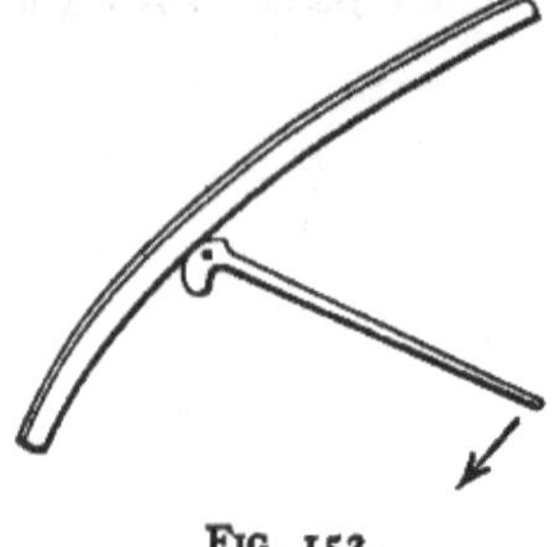

Fig. 152.

For heavy work, on the bending slab cast-iron discs about 4 inches in diameter and about 1 inch thick are used to hold the frame angles as they are being bent.

On the bottom of these discs a short lug is cast. It is located off the center of the disc, to give varying distances from center of lug to center of disc. The lug is square in section and just enough smaller than the square holes in the slab so the disc can be easily dropped into place and removed.

The lug holding the disc, with a slight variation possible in position, according to the way the disc is held as the lug is dropped into the hole in slab, can be arranged to form any curve with other discs. The hot angle frame is forced against the side of the discs and takes its shape from the curve formed by the discs, they having been previously placed to suit a set iron or mould.

This method of shaping frame bars is used for both channel bars and angle bars.

For the type of frame common in the standard merchant boat, the channel bar is generally used at certain places in a ship structure according to the design. "Web frames" are fitted where local strength is specially required. In some designs, there is a system of "deep framing" which consists in having "web frames" fairly close together with lighter framing in between but in the standard

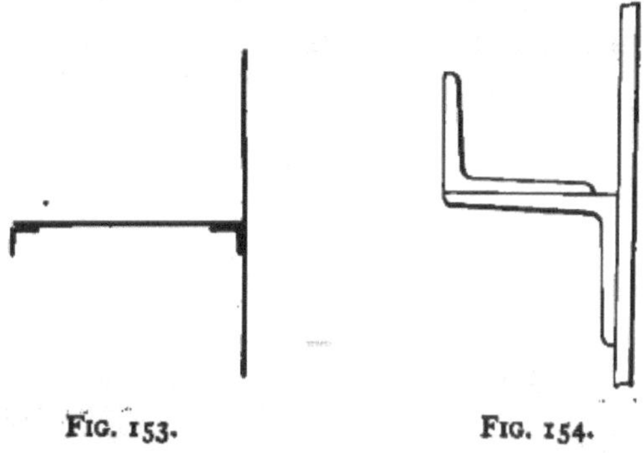

Fig. 153. Fig. 154.

type of boat, these "web frames" are not so common and the regular frames are carried for the length of the ship.

"Web frames" are made up of a plate and two angles, one angle against the shell facing the same direction as the frames and the other angle at the "inboard" edge of the plate riveted to it, and called a "stiffener." This angle can look the same direction as the other or not, as desired. The depth of these web frames varies according to the design and may be anywhere from 12 to 30 in. (Fig. 153.)

Sometimes a frame is made up of a deep angle, the narrow flange of which is against the shell plating and a smaller angle with both flanges alike is riveted at the "toe" of the deep flange, both angles being back to back with the other flange of the smaller angle pointing in the opposite direction from the flange of the larger angle which is against the shell of the ship. This smaller angle is called a "reverse frame," is used to help stiffen the frame and often is the fastening for the wood cargo "battens." (Fig. 154.)

The usual spacing for frames of a ship about 8000 tons

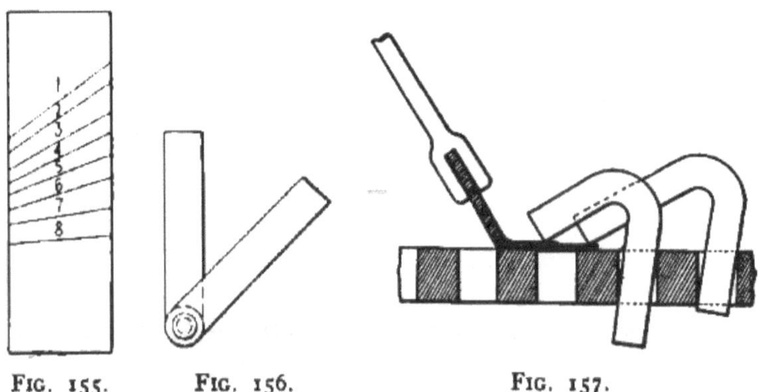

FIG. 155. FIG. 156. FIG. 157.

is between 26 and 30 in., with a little closer spacing at the ends of the ship, coming down, maybe, to 24 in.

The bevel of every frame is different and the Mold Loft supplies a "bevel board" (Fig. 155) on which is marked off the bevel of the shell flange of the frame. An adjustable bevel (Fig. 156) is used for checking up the bevel when bending the frame. An angle bar on the bending Slab with the toe against the set iron and the other flange being beveled by the bar which fits down over it and can be moved along the length to suit the different bevel required is shown in Fig. 157.

Because it is best to have an "open-bevel" of the flange of the angle against the shell plating, it is necessary to have the face of the angle "look" toward amidships and it is customary to carry this method of the flanges, facing each other to a certain point in the ship which is either directly amidships so that all the frames in the forward body look aft and all the frames in the aft body look forward; or the frames in the forward body look

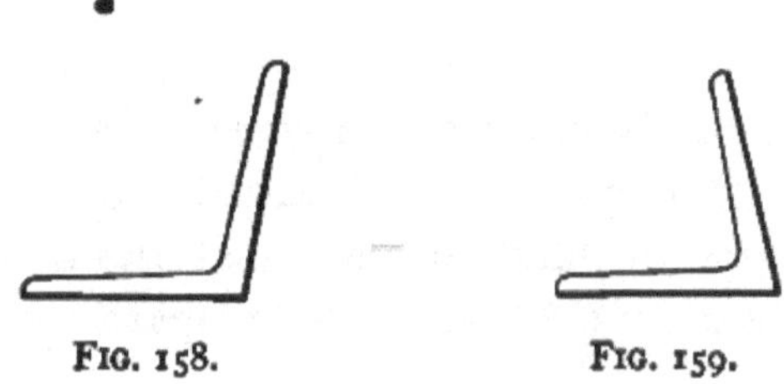

FIG. 158. FIG. 159.

aft as far as practicable in order to keep the style of riveting in the shell plating as much alike as possible.

This "open-bevel" on the angle is necessary in order to give the riveting gang an opportunity to reach the rivet. (Fig. 158.) If the bevel is a "closed" one, it is difficult to get the rivet in the rivet hole. (Fig. 159.)

It is necessary to have the ship designed in such manner that the work can be accomplished according to the equipment of the yard in which the ship is being built.

In the older type of ships, the outer plate of the tank top, called the "margin" plate, was carried out flat sufficient for the "landing," and then bent down to meet the shell plating at right angles. (Fig. 160.) Many times the frame angles were carried down through this "margin" plate into the Inner Bottom and secured to the floor.

This necessitated water-tight "staples" around the frame.

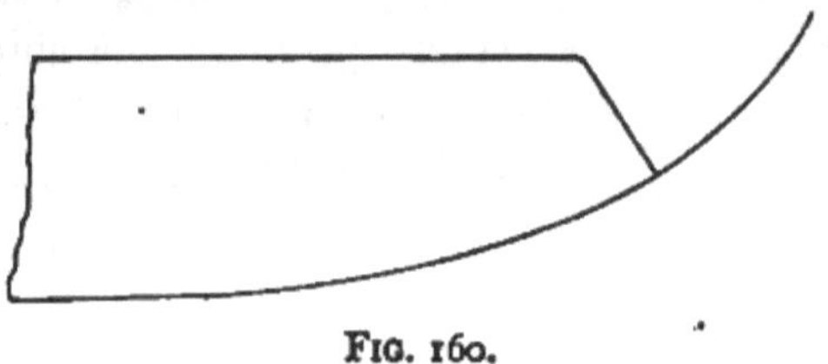

FIG. 160.

This design has been altered in some of the later boats.

In the later type of "fabricated" boats where the designer keeps in mind the fact that the different parts of a ship are often made many miles apart, the "margin"

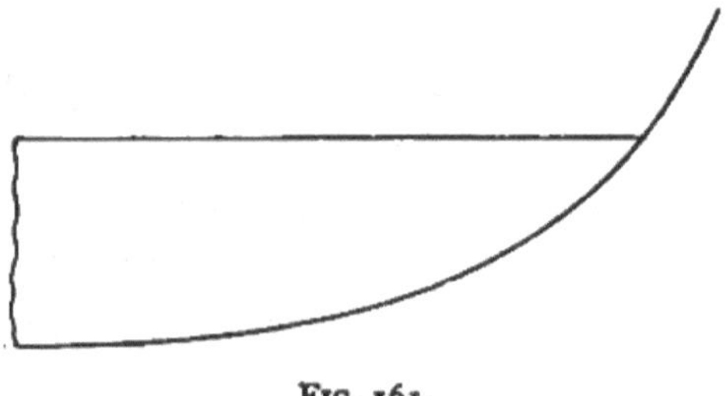

FIG. 161.

of the tank top is carried out level to the side of the shell plating (Fig. 161) and the frames come down to the tank top and are secured to it by means of a "bracket" and "clips." (The clips are short pieces of small angle bars.)

In a design of this kind, all the structure of the floors can be made up entirely independent of the framing and vice versa.

After the frames are erected, the " riggers " or " erectors " hold them in position at the upper end by means of wide, wood " ribbans " bolted to the frame, or with one strake of shell plating.

Frames are generally in one length, which varies according to their location in the ship. With the common type of " Poop," " Bridge," and " Forecastle," the frames run up to these decks or stop at the upper deck in way of the forward and aft " wells."

The stern of an ordinary merchant ship is round at the extreme end and is carried in a curve to the midship portion of the vessel. The " transom " frame is at the stern post of the ship and is usually built up of a solid plate or plates similar to a bulkhead. The "cant" frames are light framing attached to the after side of the transom frame and spread out like the ribs of a fan, being carried normal to the curve of the deck as far as possible. The " cant " frames are designed sufficiently strong to support the after " overhang " of the ship which is usually about 10 or 12 ft.; that is, the length of the ship aft of the transom frame is about this amount. (Fig. 162.)

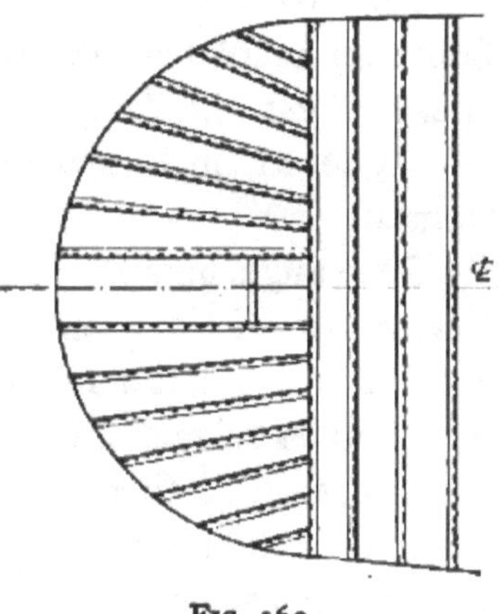

Fig. 162.

The majority of the vessels now being constructed for cargo carriers are built on the "transverse framing" plan, which has frames as just described.

Battleships and other long vessels requiring more longitudinal strength secure much of it through additional longitudinals in the Inner Bottom and other increases in strength in the stringers, side shell, etc.

A new type of framing for cargo carriers which gives more longitudinal strength than the transverse, yet allowing sufficient for all transverse strains, is called the "Isherwood System," named for the man who invented and patented the system.

This consists mainly in having deep web frames built up of plates and angles spaced about twelve feet apart along the length of the ship (instead of about twenty-seven inch spacing for twelve-inch channel-bar frames).

Deep angle bars are carried along the length of the ship, having two of them on each plate. The web frames are notched out to allow the longitudinal angles to pass them. This applies to each and every plate on all the decks, tank top and shell plating.

The depth and sizes of the web frames vary according to the size of the ship, as do the longitudinal angles on the plating.

Workmanship, so far as riveting and other work needed, is the same as for any other boat. This brief description of Isherwood Framing is given so that any men working on this type of boat will recognize it and understand that it is entirely different from the older method. Except for the difference of framing the other work of hanging plates, bolting, reaming, riveting, etc., is the same.

An outline of an "Isherwood" framed ship, showing the longitudinal bars against the deck, shell and tank top plating is shown in Fig. 163. The web frames are spaced at intervals, indicated by dot and dash line.

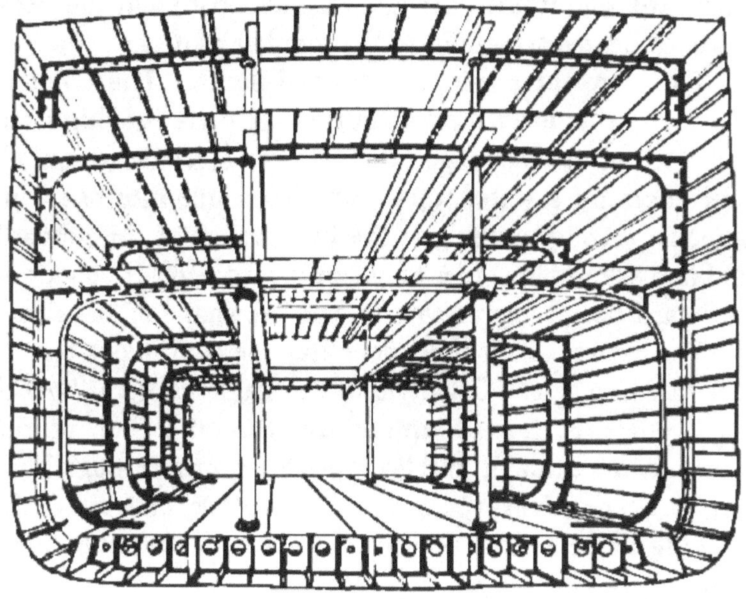

Fig. 163

CHAPTER VIII

Floors

Directly in line with each frame there are floors built up of "girders," the depth being the full height of the Inner Bottom and extending in length from the Vertical Keelson to the side of the ship at the bilge. The floor is the part of the continuous transverse framing of the ship which is made up of two floors, one port and one starboard, one frame port and one starboard and the deck beam or beams according to the number of decks.

There are three common types of floors. The "water-tight" floor is formed by a continuous plate running from the keelson to the bilge of the ship, surrounded on all four edges by a water-tight angle bar made into one continuous piece by welding at the two corners of the inboard end and the upper corner of the outboard end. (The lower outboard part being curved to suit the shape of the ship.)

The water- or oil-tight floor has the continuous bar on the caulking side. Usually another angle called a "backing" angle is carried around on the other side of the floor, rivets through the floor plate passing through both angles. The rivet spacing on the other flange of the backing angle is not so close, as it is non-water-tight. (Fig. 164.)

Another type is called a "solid, lightened" floor and

consists of a solid plate extending from the keelson to the bilge of the ship with holes cut in the plate. These holes lighten the weight of the plate by removing some of the material and also serve as access holes for workmen moving around in the Inner Bottom. In this type the angle is

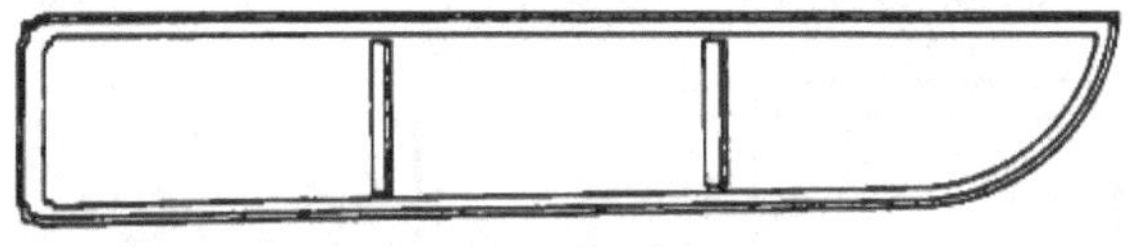

FIG. 164.

not continuous, being cut at inboard end for keelson bars. (Fig. 165.)

The third type is called a "bracket" floor and is formed by a heavy angle with a flange about 3½ ins. against the shell plating and the other flange about 7 or 8 ins. deep. The angle against the tank top of the same size, connected at the inboard end with a bracket carrying one angle

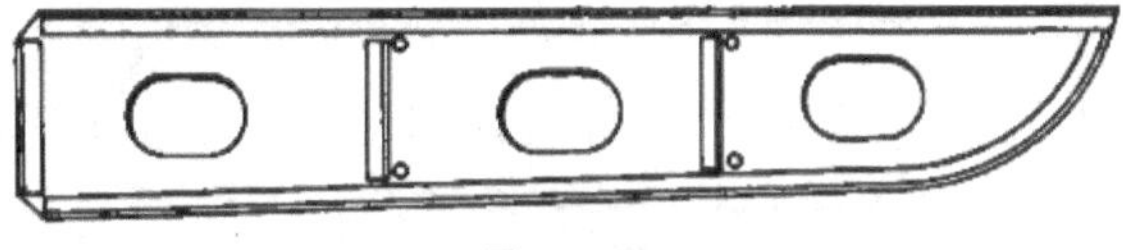

FIG. 165.

against the keelson and extending outboard about 2 or 3 ft. and stiffened at the outboard end by means of a small angle. The outboard ends of the shell and tank top angle are connected in the same way by a smaller bracket plate. Both of these plates are lightened by a circular hole. The mid part of this type of floor is held together by deep angles placed in line with the intercostal longitudinals, to which they are riveted after the floors and longitudinals are in place. (Fig. 166.)

The water-tight floors form the boundary of the different compartments in the Inner Bottom and are made water-tight or oil-tight according to the design.

It is customary to locate a solid floor at every third

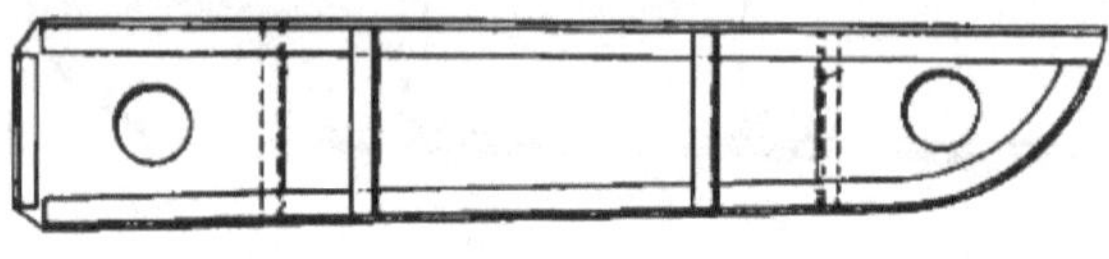

Fig. 166.

frame space with two bracket floors in between, except under engine room when all floors are solid, lightened, for additional, local strength.

The angle on the floor faces in the same direction as the frame angles against the shell plating.

The solid, lightened floors have an air vent at the upper

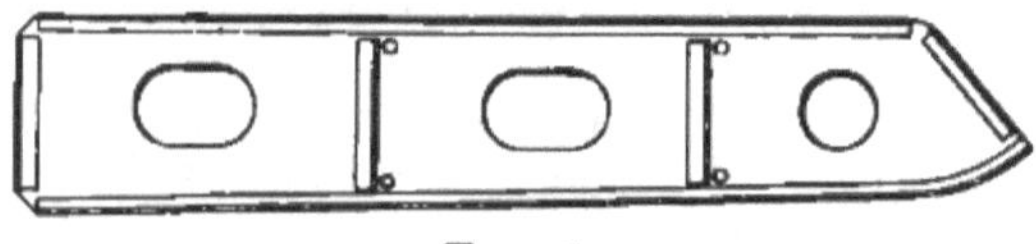

Fig. 167.

part and "limber" holes at the lower part. (These limber holes are used for passing through a "limber chain" which can be drawn a little forward and a little back, if necessary, to give free opening in case of stoppage, for draining either water or oil toward the suction pipe.)

Fig. 167 illustrates the floor of a ship having the margin plate turned down to meet the bilge, while the floors in

Figs. 164, 165 and 166 are for tank tops carried level out to the ship's side.

When erecting floors the Piping Department should be consulted, as they have to install their pipes for the " Ballast piping " system at the same time the floors are going in, as much of their piping is in such lengths that it can go into place only when *certain floors* are in the ship.

Longitudinals

There are usually two main longitudinals running from near the bow as far aft as the end of the Inner Bottom. In addition to these, there are local longitudinals under the Engine Foundation and three or four short longitudinals near the bow.

Longitudinals are designed primarily to reinforce the strength of a boat in a fore and aft direction. They are built in the Inner Bottom and are secured to the shell plating, tank top and floors; thus tying the whole part of that portion of the ship together. Taken in connection with the keelson, there are 5 main girders about 4 ft. deep (the depth of the Inner Bottom) which tend to resist any bending of the bottom of the boat.

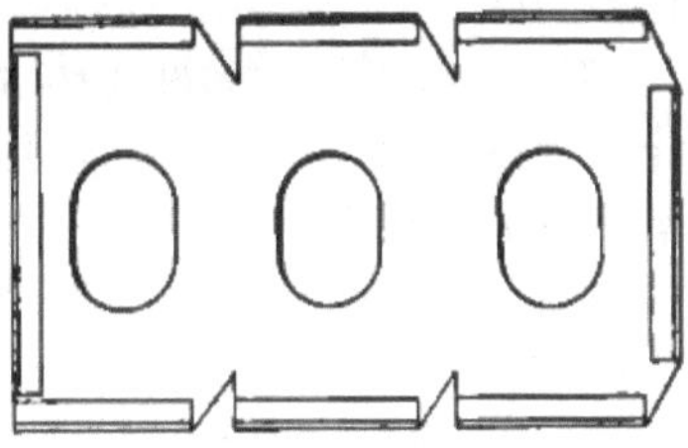

Fig. 168 shows a longitudinal made to extend between four floors.

Fig. 168.

Longitudinals are divided as continuous and intercostal, water-tight and non-water-tight, but in the case of a merchant steamer, the longitudinals are intercostal and non-water-tight, allowing the compartment to extend from the center line keelson out to the side of the ship.

Any structure on the ship which must be fitted in short lengths is called " intercostal " work and as these longitudinals are fitted between the solid floors which are on every third frame space they are called " intercostal longitudinals " or for short " intercostals."

As the floors under the Engine Room are all solid,

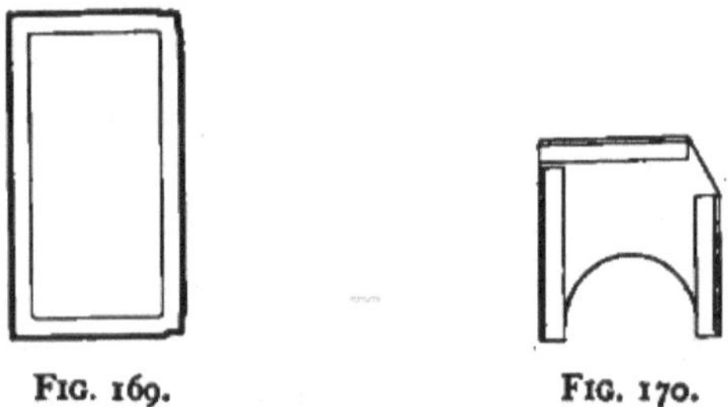

Fig. 169. Fig. 170.

these longitudinals are fitted between each floor in that case.

A solid longitudinal between two floors is shown in Fig. 169.

The longitudinals are designed with access holes so that the workmen may travel unhindered to any part of the Inner Bottom.

Fig. 170 shows a half longitudinal between two floors. This does not extend down to the shell plating but is used to stiffen the tank top and tie the floors together.

After the flat keels and bottom plating have been laid and the vertical keelson has been bolted (and riveted if time and opportunity permit) the floors are swung into place, and lined up, the rivet holes in the bottom plating giving the location for the flange of the shell angle of the bottom of the floor. These floors are securely bolted in place, care having been taken to place stop-waters (usually of lamp wick) between the keelson and floor angles of all water-tight floors.

The intercostals are now lowered into place, care being taken to ascertain that they make a good tight fit between the floors.

The longitudinal strength is increased by one or more

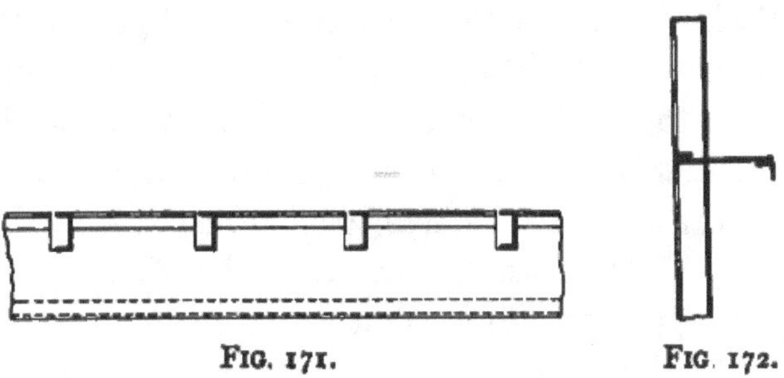

FIG. 171. FIG. 172.

" side stringers," consisting of a narrow plate, notched out over the frames, with a continuous angle on the inner edge and intercostal shell clips (between the frames) on the outer edge. There are usually two such stringers between the tank top and lowest deck. They are secured to the transverse bulkhead on both sides by means of heavy brackets, thus continuing the strength of the stringer.

Fig. 171 shows a plan view of a side stringer.

Fig. 172 shows a section of a side stringer.

CHAPTER IX

Tank Top, Inner Bottom and Peak Tanks

The bottom of the modern steamship is built up by an inner and outer bottom, so-called. This is formed by the outer shell plating and the "Tank Top." This Tank Top is really the lowest deck, about 4 ft. above the bottom plating. The space between the Tank Top and the shell is the Inner Bottom.

The Tank Top is about the same thickness as the steel deck above and is riveted for oil-tight work. At certain places where necessary for access "Manholes" are cut in the plating just large enough to allow the passage of a man's body. A common size is 12 ins. by 18 ins. in the clear opening. These are "Flush" or "Raised," according to the location in the ship. Those being in the cargo space are flush while those in the Boiler Room are raised. The "raised" part being in the angle coaming which sets the cover up from the Tank Top plating.

Plating of the Tank Top is carried the same as for any of the decks; being laid out in strakes, with double riveted seams and butt-laps. The center line plate, directly over the keelson is called the "Rider Plate" and is always an "inside" strake, as it is the first down and because the shoring for deck beams above can be put in place as soon as needed, before work on any of the other strakes is done.

The strake against the shell plating is called the " Margin Strake." Sometimes this strake is turned down just after clearing the seam, at right angles with the turn of the bilge plating (Fig. 173), but in some boats this

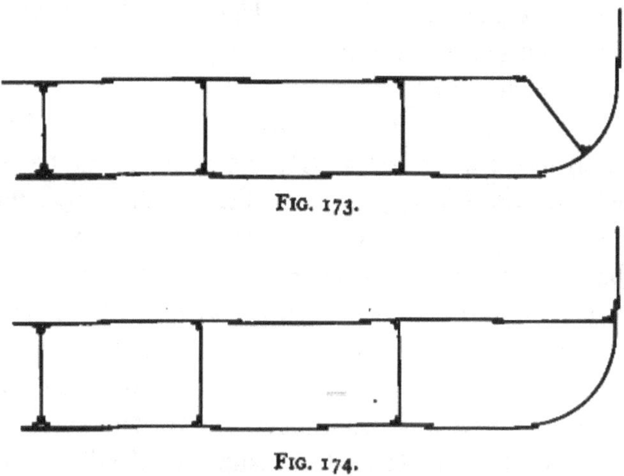

Fig. 173.

Fig. 174.

plate is carried out level to the side of the ship. (Fig. 174.)

In the " Cargo Holds " the Tank Top is the flooring on which the cargo is supported.

The " Inner Bottom " is formed by the Tank Top and bottom of the Shell Plating and in a coal-burning ship: it is used for carrying water ballast when the ship is traveling " light " (with no cargo), while in an oil-burning vessel the space is devoted to carrying oil as fuel for the boilers.

The oil-tight or water-tight floors at certain intervals form the different compartments of the Inner Bottom. It is possible for a workman to travel all through the Inner Bottom by means of the Manholes in the Tank Top and bracket or open floors. These, together with the solid, lightened (access holes) floors form the third type of floor in the Inner Bottom.

When a workman is traveling around in the Inner Bottom with a candle for a light he is likely to become confused and lose his way back to the Manhole by which he entered, especially so when working on a large vessel like a battleship, but on a smaller, cargo boat this is not so apt to happen. If he is working with a "portable" electric light the cord will pilot him back to the entrance.

An Inner Bottom of this type is a partial safeguard against sinking as a hole might be stove in the bottom of any of the compartments and the ship would still be able to float. Some ships would still float if the hole happened to come on a water-tight floor, when two compartments would be "flooded." This depends on the design of the ship. Because of this possibility, battleships have the Inner Bottom carried up the side as far as the armor plate and there are six or seven longitudinals, some of them being water-tight to reduce the area of flooding and consequent loss of buoyancy. Ships of the cargo carrying type use the Inner Bottom for fuel oil or water ballast.

The Peak Tanks are two built-in tanks used for "Trimming Ship" and partly for another purpose.

All sea-going vessels are required to have a "Col lision Bulkhead" in the forward end, sufficiently far from the stem to give a good-sized compartment.

This is necessary, because if the ship is in collision with another and has the bow stove in or from other causes springs a leak this bulkhead will serve to shut off that part of the ship so the water which enters will be confined to one place and not sink the vessel. To make this bulkhead as tight and strong as possible to serve its pur-

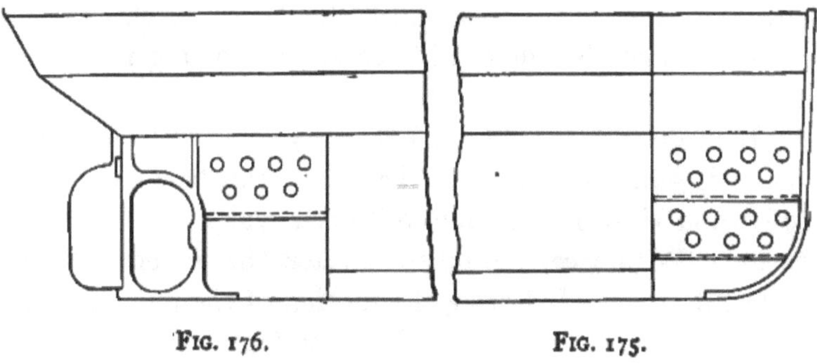

Fig. 176. Fig. 175.

pose, it is well braced with stiffening angles and the plating itself is amply thick to help withstand any bending due to water pressure.

This space between the Collision Bulkhead and the stem of the ship is called the "Fore Peak." (Fig. 175.) There is another Bulkhead, similar in design at the after end of the ship, located a short distance ahead of the stern frame which forms another compartment, called the "After Peak." (Fig. 176.)

As the Collision Bulkhead is there partly to meet an

emergency, it follows that the Fore Peak is generally empty and dry. Both of the Peak Tanks are connected up with the piping system so they can be flooded or emptied as desired.

When a ship has the cargo stowed so that she does not set on an "even keel" (horizontal) it is customary to partly fill one of these tanks to load down one end of the ship so the uneven condition will be offset by the added weight of the water. This is called "trimming ship." When the ship is "down by the stern" then water is run into the Fore Peak until she trims level. If the ship is "down by the head," water is run into the After Peak.

Because it is not always necessary to entirely fill these tanks it will often happen that they will be only partly filled and then a condition will exist which must be corrected. With free, unbroken surface the water will roll from side to side following the motion of the hull and this will tend to accentuate the rolling motion of the ship. To avoid this, "Swash Plates" are fitted in the tanks, running in a fore and aft direction, to break up the side roll of the surface of the water. These plates are fitted up free from each other, with large holes cut out of them. The plates and holes are "staggered" so they form a baffle to side rolling. By using these swash plates any amount of water can be let into the tanks without having any trouble from them.

If the ship springs a leak while there is water in the Fore Peak the water which leaks in will only join the water already there and will go no further, being held within the Fore Peak by the water-tight bulkhead and thus still save the ship.

The piping is connected to the pump in the Engine Room so that any water that is in the compartment can be pumped out quickly, but if the leak is a bad one the water is left there until the ship makes port and can be put into dry dock for repairs.

CHAPTER X

Stem, Stern Post and Rudder

The stem of the modern cargo boat is a steel casting, in one or two pieces, about 2¾ ins. by 10 ins. of the same section throughout its length. The lower part, as it rounds down to pass under the boat, is called the "Forefoot." This steel bar is set after the vertical keelson is in place and part of the shell plating is on and riveted. The two lower bars of the keelson are spread apart sufficiently to take the width of the stem and it is fitted neatly in between these bars and riveted to them.

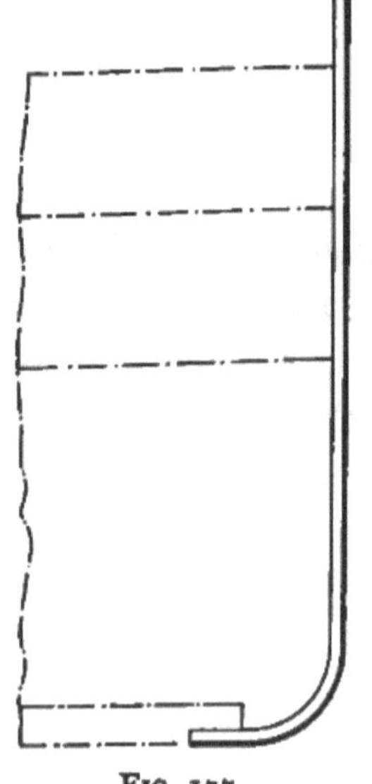

Fig. 177.

Before placing the stem casting the positions of a water line and a frame station are marked off on it by means of a wood template lifted from the Mold Loft floor, so that later the erectors will have something from which they may measure in getting the correct location to set the stem at the correct height and rake.

After the rivet holes are laid off and drilled and the

stem is ready to set up, it is raised in place by a crane and held there while bolts are passed through the bottom keelson bars and side shores are carried half way up on each side, catching angle clips which are bolted to the stem through some of the rivet holes. The side shores having the appearance of an inverted letter Λ. A shore on the center line, in front, will take the strain in that direction. The ship carpenters now check the position of the stem and regulate it so that when the forward plates of the shell are fitted to it the rivet holes will come in line and be ready for bolting up.

The upper end of the stem is called the "Stem Head" and is carried up above the Forecastle Deck in order to fasten it to the sheer strake of that deck or bow chock (sometimes called "apron plate"). In most ships this stem is straight above the curve at the Forefoot and is "raked" forward about 1 ft. to overcome the optical illusion which would make it look as though it was tilting backward, due to the curve of the "sheer" or outline of the upper part of the profile of the boat.

The shell plating is lapped onto the side of the stem and heavy rivets are driven clear through with a countersunk head and point so that both sides of the ship are flush.

The stem is reinforced by means of "Breasthooks" which are bracket plates directly behind the stem, carried out to the side of the ship and aft to the Collision Bulkhead, according to the shape of the bow of the vessel. These breasthooks are placed horizontally and are connected to the side of the ship by a stringer built up of a plate and angle which runs aft in order to distribute the load along the side if the stem should ever suffer a blow. There are two or more of the breasthooks according to the design as laid out in the drafting room, and they are placed about midway between the decks. The plates of the breasthooks are the thickness of the average deck plate. (Fig. 178.) These Breasthooks also serve as "Panting Stringers."

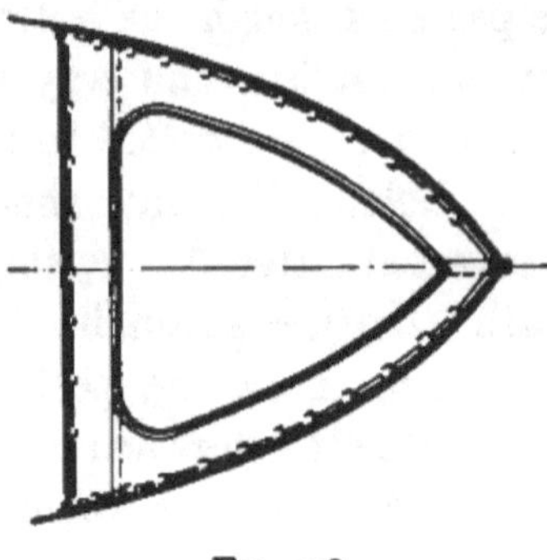

FIG. 178.

The Stern Post, or Stern Frame, of a merchant steamer varies with conditions under which she is built, as to the desirability of using a large casting and building the plating onto it or providing a smaller one and using a built up section of steel plates to continue the ending of the shell plating.

The first consideration which is taken up in a new design is the question of one or two propellers. As the majority of cargo boats, having a slow speed, are fitted with only one propeller, that is the kind in which the description of the Standard Cargo Steamer is most interested.

In passing, it is well to note that in twin-screw ships the two shafts from the engines to the propellers are lined parallel with center line of the ship. Where the shaft emerges from the shell plating, in the counter of the ship, a boss is fitted around which the shell plating is worked and a stuffing box is placed inside to preserve water-tightness. The after end of the shaft, just forward of the twin propeller, is supported by means of the shaft strut, having the upper arm carried up, on an angle from the horizontal, to meet the counter of the ship and the lower arm carried down near the keel. These struts end in a broad "palm" which is well riveted to the hull structure. The struts are made oval in section to pass through the water with as little disturbance as possible.

The stern frame of the single propeller (or "screw") steamer is commonly formed by a steel casting in one piece, consisting of the "Stern Post" and the "aperture" for the propeller, and the "Rudder Post." The Rudder Post is oblong in section, designed to pass easily through the water on the forward side and to carry the weight of the rudder on the aft side. The aperture is sufficiently large to allow room for the propeller blades to swing with ample clearance both for height and also between the propeller cap and Rudder Post.

The frame forward of the aperture is made long enough, fore and aft, to take the ends of the shell plating which run onto it, with a "boss" or larger, circular part where the propeller shaft comes through from inside the ship. The opening of the aperture is curved. The forward part is a large curve with a short curve at the top and a nearly straight part at the bottom. The stern post extends above the aperture and an arm is provided to extend along the keel to allow for a connection with the plating.

The after edge of the Rudder Post is "plumb" and square; and is fitted with lugs of square section, called "gudgeons," through which a vertical hole is drilled to take a "pintle" of the rudder. (Fig. 179.)

Another type of Stern Frame is made up of a small casting and U-shaped plates, but it is not so strong and

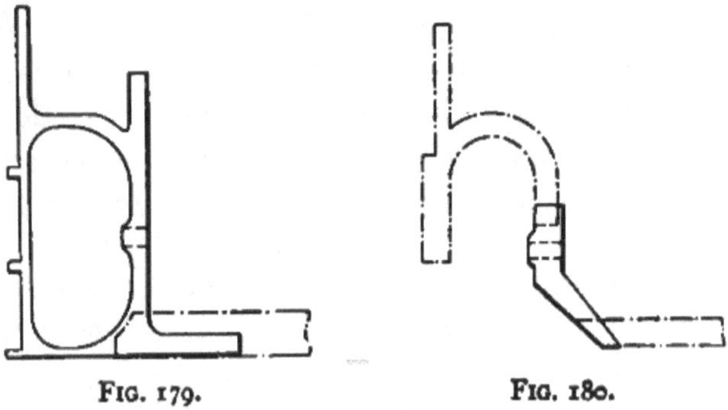

Fig. 179. Fig. 180.

there is considerable shipfitting to get it in line. The only advantage in a frame of this kind (Fig. 180) is in the small casting required.

The Rudder varies in shape and structure according to the ideas of the designer, but is generally made up of a "Rudder Stock" which is rounded on the forward side, has arms extending aft to take the rudder "blade," and extends up into the "rudder trunk" inside of the ship. In some designs the "arms" are shaped to take a single rudder blade or plate and sometimes the rudder has plates on both sides of the arms and is filled in between the plates and arms, solid, with wood or other material. On the forward side of the Rudder Stock, straight hooks are cast and shaped to pass down through the gudgeons on the after side of the Rudder Post. These hooks are called " pintles " and are to take the weight and strain of the rudder. (Fig. 181.)

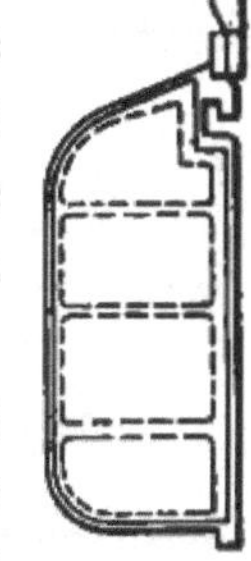

Fig. 181.

On the outside of the rudder stock there are lugs which are designed to meet others on the rudder post when the rudder is at the maximum swing. This is done to relieve the strain on the steering gear, in part, and to act as a " positive stop."

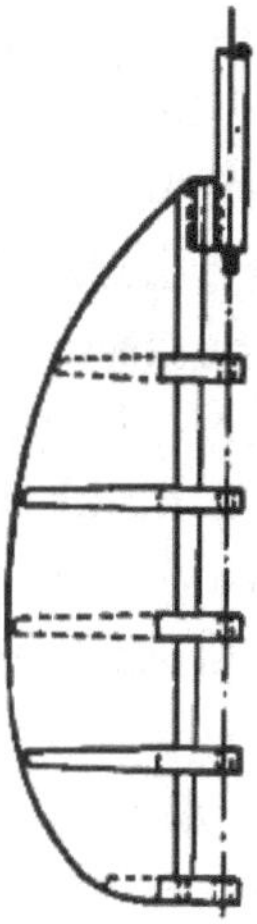

The Rudder Trunk is a box built into the hull, large enough fore and aft to allow " shipping " of the rudder stock, and is water-tight to prevent leaking of water up into the hull when the stern is deep in a " following sea." It is made of plate and angle, is caulked tight, and extends up to a sufficient height above the waterline.

Fig. 182.

The upper end of the Rudder Stock is formed to take a yoke which is operated by the steering gear. When the yoke is swung to one side or the other the whole rudder is turned by means of the rudder stock.

Many merchant vessels have a "plate" rudder. This is made by using a steel casting for the rudder stock with flat arms on alternate sides of the plate, to which it is fastened. (Fig. 182.) This type of rudder is light in weight and not difficult to fit into place.

CHAPTER XI

Bulkheads and Hatches

Bulkheads are vertical walls dividing the ship into rooms called "compartments." They are divided into longitudinal and transverse bulkheads. These being either water-tight or non-water-tight.

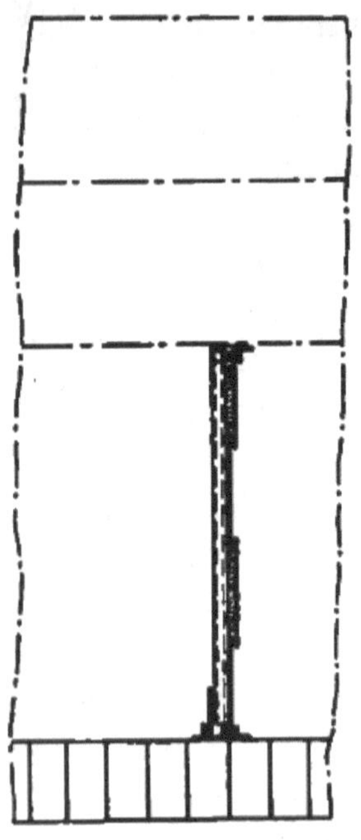

FIG. 183.

The main bulkheads of the ship are transverse and water-tight up to the Lower Deck. The water-tight bulkheads are designed to withstand the bending which will tend to distort them if the compartment is filled with water. To prevent this bending, the bulkhead plating is built on a heavy framework of either angle or channel bars, in most places the latter. (Fig. 183.) These bars are called "stiffeners" and extend vertically from the top to bottom of the bulkhead, being secured to the deck above and at the bottom to the Tank Top by means of clips or sometimes by a heavy bracket. (Fig. 184.)

In order to connect the edges of the bulkhead, a bar

is carried around and riveted to the bulkhead and adjacent plating; that is, Tank Top, side shell and deck above. This bar is called the "Boundary Bar" and in case of the

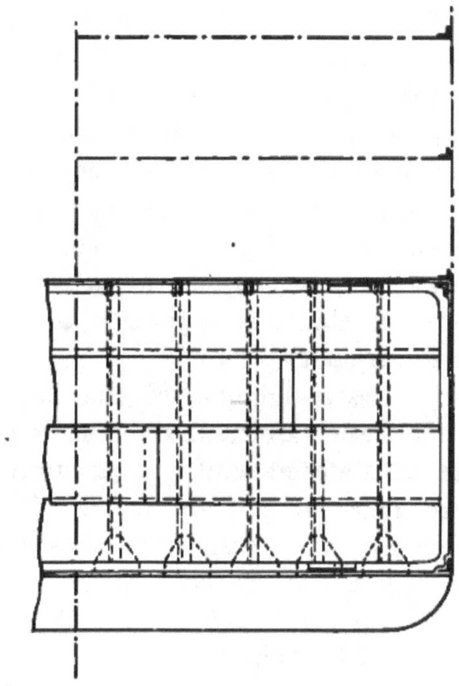

FIG. 184.

water-tight bulkheads, it is the bar which is caulked. The bottom and side edges are often reinforced by having another angle bar on the back side of this plating. This is called the "backing" bar.

In yards where the service of an over-head crane is available, the bulkheads are built on the ground just clear of the bow of the ship. The stiffeners are laid horizontally on trestle work and the plating is placed, bolted and riveted so the whole bulkhead of plating and stiffeners are in one piece. The bars are riveted on and then the bulkhead is lifted whole by the crane, carried over the ship and put into place.

In those yards where it is not practical to build the bulkheads on the ground in front of the shipway, the bulkheads are built directly on the ship. In this case, the vertical stiffeners are set in position, secured at top and bottom and the plating hung in place. Care being taken to prevent any sagging of the plating due to loose bolting. After the plating is in position, the backing and boundary angles are put up.

To preserve water-tightness of the bulkheads, the edges of all the plating are caulked. This is done on the side opposite the stiffeners as it is impractical to carry a caulking edge under the stiffener angles. The caulking bar (Boundary Bar) of a water-tight bulkhead is " faced " in the same direction as the frame angles. This is done to allow an " open bevel " when the bulkhead happens to be at the end of the ship.

When a main, water-tight bulkhead, without any door allowing passage through it, is in course of erection, it is customary to cut an "access" hole near the Tank Top of sufficient size to allow the passage of a man. This hole is covered with a " patch " plate after the work in that part of the ship is complete. This is to allow free passage from one compartment of the ship to another without the necessity of climbing up over the top of the bulkhead and down the other side. When the design calls for a water-tight door to be installed, the opening in the plate for this door is used and no other access hole is cut.

Because the lower part of the bulkhead will have a greater pressure due to the " head " of water in the compartment if the ship is leaking, the plating of the lower strakes is about one-eighth of an inch thicker than the top plates where the pressure is less. It is for this reason, also, that the stiffeners have an extra fastening at the lower end. It is customary to arrange the plating of the main bulkheads which have to take a water test with double riveted seams, and lapped butts. Care should be taken to lay off the plating so that the seams will not come too close to any " flat " or deck where the riveting would foul the deck angle.

The **Engine Hatch** consists of an opening in each of the decks directly over the engine. This is enclosed by a light bulkhead carried around all four sides from the lower to the highest deck which may be the Bridge Deck. Here the opening is covered by means of an Engine Room Skylight which is made portable so that it can be lifted off by means of a crane if there is any repair work to be done to the engine. On the top of the skylight, small shutters (or covers) are fitted so that they will hinge up on end and give plenty of ventilation to the engine room. For additional lighting, when the covers are closed, they are fitted with glass ports, or windows.

In order to obtain access to the Engine Room, a door is cut in this enclosure and a ladder is carried from there down to the Engine Room floor. This door is usually on the Upper Deck in the bridge enclosure where it is customary for the engineering force to be located.

The **Boiler Hatch** is another opening similar to the one over the engine in that it goes up to the weather deck and is enclosed on all four sides below that deck. An opening through the enclosure gives access to the Boiler Room by means of a ladder. The top of this hatch is covered with what is termed the "Fidley Hatch," a square boxlike arrangement standing a little above the deck, which allows the funnel to pass up through. It is covered with grating to give sufficient outlet for air from the Boiler Room.

The **Cargo Hatches** are sufficiently large to allow ample room for handling the cargo in and out by means of the Cargo Booms.

Deck beams are cut in way of the hatches and are secured to fore-and-aft hatch girders, below the deck. The deck beams at the ends of the hatches are of extra heavy design and depth to take the additional strain off the ends of the fore-and-aft hatch girders. Some ships have the second beam beyond also of a heavier type than the regular deck beams, in order to help reinforce the framing at these places.

The fore-and-aft hatch girders of the Lower Deck are sufficiently deep to take the ends of the Hatch Strongbacks. These are generally I-beams which are portable and rest in sockets of angle iron riveted to the girders. An angle iron is carried around the Cargo Hatch, at the Lower Deck, set back from the edge of the hatch a few inches for a landing for the Hatch Cover.

Around the Cargo Hatches, on the Upper (weather))eck a Hatch Coaming of plate stands above the deck bout 3 feet, generally in the same vertical line with the eams at end of hatch and the fore-and-aft girder.

This coaming is braced strongly with bracket plates ɔ the deck and carries a small angle around the inside, ɜt down from the top a few inches, as a landing for the Iatch Cover.

Angle-iron sockets to receive the Hatch Strongbacks re secured to the Hatch Coaming to allow the tops of the -beams to come in line with the Hatch Cover angle on ıside of the coaming.

Cargo Hatch Covers for the design just described are enerally of wood plank about 3 inches thick arranged in ɜngths and widths to completely cover the hatch opening.)ver the Hatch Cover, on the weather deck, canvas arpaulins are laid and secured tightly by carrying over he edge of the coaming and held against the coaming by neans of Battens.

Some ships have a steel Hatch Cover hinged on either he forward or after-end. This is built up of plates and ngles, stiffened to prevent buckling and made water-ight by means of a rubber gasket on the under side, ıear the edge. The cover is held down by "dogs" (clamps).

These steel Cargo Hatch Covers are heavy and require nechanical means to open them. Often a steel post is ocated near the end of the hatch and tackles lead to the op of it or else to a nearby bulkhead.

The **Shaft Alley** or Shaft Tunnel, is a long, narrow

passage from the after end of the Engine Room to the after Peak Bulkhead. It is built around the " tunnel shafting " and protects the shaft from the cargo in the after Cargo Holds. The plating is well stiffened, on the inside, to withstand the pressure from the cargo and it is caulked water-tight on the outside to prevent flooding of the Engine Room in case of a leak in one of the Cargo Holds.

The sides of the Alley are vertical. The top is sometimes built flat and level across and sometimes it is curved. The beam stanchions at the ends of the cargo hatches rest on top of this Alley and it must be strengthened locally to carry the weight down onto the Tank Top. As the shafting for the propeller is on the center line of the ship, the shaft alley is built off center, giving ample passage for one person between the shaft and the starboard side of the Alley, as the shaft must be carefully watched and the bearings oiled. (Fig. 185.)

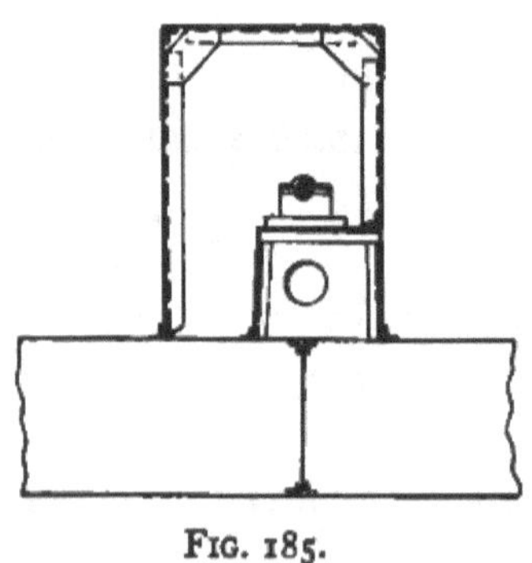

Fig. 185.

Shaft Stools or Foundations, are really steel boxes which are built to take the bearings of the different parts of the shaft, which is made up of about five lengths in the average size ship. These stools are built against the port side of the Shaft Alley, and rest on the Tank Top.

CHAPTER XII

Stop Waters

Riveted joints may be made tight without caulking by placing tarry material between the faying surfaces. This material may be made of strips of flannel or lamp wick dipped in red lead. Felt sheeting is also used for this purpose. At some places a kind of putty known as

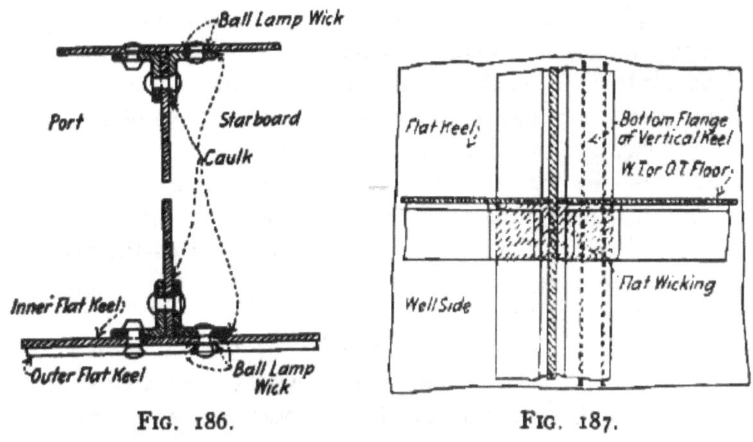

FIG. 186.

FIG. 187.

Vulcan Cement is used. This is mixed with linseed oil, to give it the consistency of a stiff dough, and worked into the joints with a case knife or small trowel. "Ruberoid" is used as soft packing. It is placed between the faying surfaces when erecting and the heat of the rivets when riveting makes it soft and sticky, thus assuring watertightness. This soft packing is used on all joints that cannot be properly caulked, also in such joints as may be very difficult to caulk, due to the complicated construction of the hull. It is used on all water-tight collar angles, tank divisions at the bottom, margin plate lugs, bars on top of tank, etc. Soft packing is not as desirable as metal caulking, but there are places in the ship where it must be used.

In general, stop-waters, oil-stops and red lead putty are means of stopping seams occurring within a built-up member, or to fill in slight irregularities of the surface of contact members, but they should never be used instead of good caulking, nor to supplement poor caulking. Their use requires care and discrimination, because too many stops encourage poor caulking and their omission, in some places, causes leaks difficult to remedy and necessitates the use of the putty gun.

Stop-waters and oil-stops are required:

(*a*) Where the work is inaccessible or difficult to caulk during testing, such as, under the outboard end of Compensating Bracket Clips to bulkheads under pressure. (Fig. 191.)

(*b*) Where the edge distance is not enough to permit chipping and caulking. Minimum edge after chipping to be $1\frac{1}{8}$ inch for $\frac{3}{4}$-inch rivets and $1\frac{1}{4}$ inch for $\frac{7}{8}$-rivets.

(*c*) Where several thicknesses of metal prevents the work being drawn up tight, and the rivet shank cannot be made to swell tight, such as bulkhead boundary angles on the Tank Top over a floor.

(*d*) At stapled corners where the material does not fit ıp tight, as at stapled corners of oil and water-tight floors. Fig. 189.)

(*e*) At Joints of warped furnace plates that cannot be

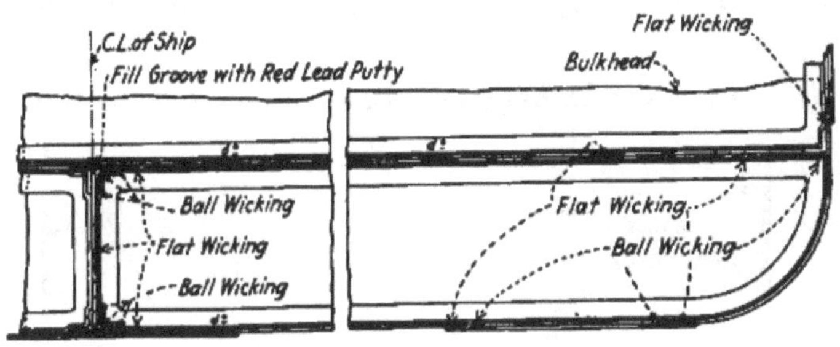

FIG. 188.

lrawn up tight, such as under shell butt strap on the bilge trake at the molded ends.

(*f*) Where oil or water tightness is desired and the

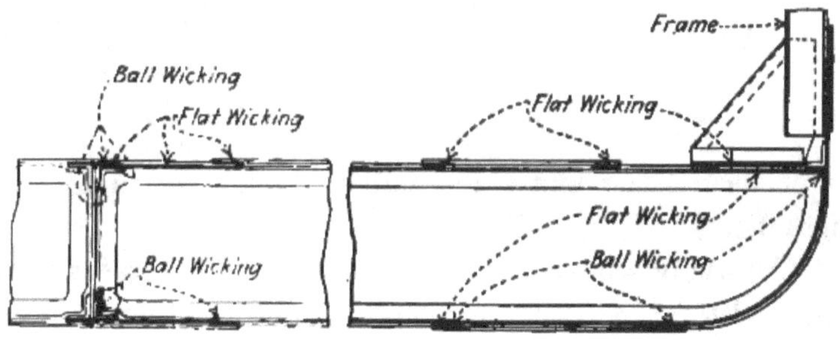

FIG. 189.

:ivet spacing is too wide for caulking, as under the Bilge ınd Ballast Pump Foundations.

(*g*) To prevent leakage around the shank of pan or

button head rivets, as found back of the shell leg of bulkhead boundary angles of bulkheads under pressure. (Fig. 191.)

(*h*) To prevent leakage from one seam to another through a rivet hole, as happens between shell seams, or Inner Bottom seams to the contact surface of oil-tight floors. (Fig. 188.)

(*i*) To prevent leakage longitudinally in a seam or joint

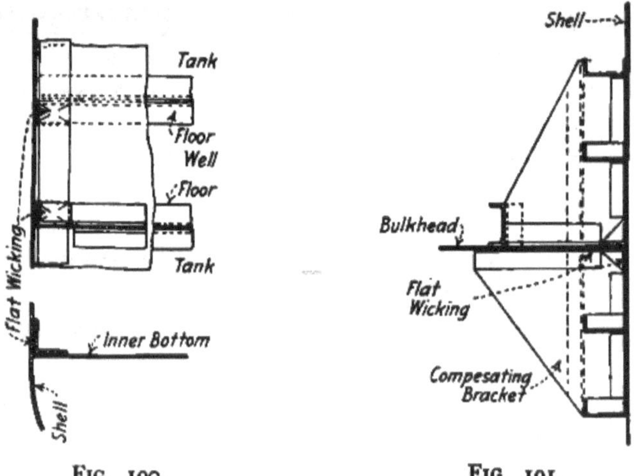

FIG. 190. FIG. 191.

as between Inner and Outer flat keel plates or under oil-tight floors. (Fig. 188.)

The use of stop-waters and oil-stops in some cases does away with the necessity for caulking and in others it does not.

Where stops are used to prevent leakage from around the rivet shank out through a seam, as under frame clips on the Tank Top, the clips need not be caulked unless a leak develops during testing.

Where stops are used to prevent leakage through a seam, such as between floor angle and shell, the edge should be caulked.

Water-tight Faying Surfaces should be coated with red lead and **Oil-tight Faying Surfaces** should be coated with a mixture of pine tar and shellac immediately preceding erection.

Stop-waters are made of flat wicking; canvas or ball wicking soaked in linseed oil, or they may be made of tarred felt.

Oil-stops are made of flat wicking, canvas or ball wicking soaked in a mixture of pine tar and shellac, or a mixture of one part red lead and three parts shellac, or they may be made of tarred felt.

Flat, water and oil stops should be ½ inch wider than the contact surfaces.

Holes in water and oil stops should be cut and not burned.

The materials used for soaking oil and water stops must be fresh and the work over the stops should be riveted up before the material has hardened.

Small holes at the corners and joints are made tight by inserting metal caulking pieces that approximate the shape of the hole and of such shape that they will not move during caulking.

Butts and edges of plates and butts of angles in strapped joints should fit closely together after being riveted. Caulking pieces are not usually inserted in any such joints in water-tight or oil-tight work.

Where leaks occur during testing which cannot be remedied by caulking or tightening up of the rivets, the leak may be stopped by the use of a putty gun, with red lead putty.

Red lead putty may be used to fill in grooves between contact surfaces before assembling, as for filling in between the top flange angles of the keel where stop-waters cross it. (Fig. 188.)

For a transverse framed steamer like the one illustrated on the folding plate in back of the book the following methods of applying stop-waters and oil-stops are good practice.

The floors, longitudinals and keelson that bound the fuel and Feed Water Tanks and the Wells below the Inner Bottom are made water or oil tight.

Keelson. Two lines of ball lamp wicking are placed, longitudinally, between the inner and outer flat keel along the entire length of the two keels and along the top and bottom wherever water-stops or oil-stops are required as shown in Fig. 186. Flat wicking is placed transversely at top and bottom flanges where necessary as indicated in Figs. 187 and 188.

Short pieces of flat wicking are placed in the seams between the shell plating opposite oil and water-tight floors and boundary angles of bulkheads under pressure. (Figs. 188 and 189.)

Water and Oil-tight Floors. Flat wicking is placed around the inboard end against the vertical keel and three feet outboard at top and bottom, as shown in Fig. 189. It is also run at the outboard end on top under the frame clips and twelve inches down around the corner. A short piece of ball wicking is placed on the floor at all stapled corners and joints between plates and liners.

Where there is a bulkhead boundary angle above as in Fig. 188, or connections that make more than three thicknesses of metal flat wicking is run the full length of the floor on top.

At the molded ends, the wicking is run at the outboard end around the curve of the bilge.

A short strip of flat wicking is also put in the seam between the shell plates under the oil and water-tight floors and a similar piece in the seam between the Inner Bottom plates over oil and water-tight floors.

Longitudinals. Flat wicking is used between the Longitudinal clips and the adjoining floor plates in the wells. For tight longitudinals flat wicking is placed all around top, bottom and sides, with ball wicking at corners and staples.

Water- and Oil-Stops on the Tank Top. A short piece of flat wicking is placed between the shell and the boundary angle and between the Inner Bottom plates and the boundary angle, over all Wells. (Fig. 190.)

Flat wicking is used under frame clips as shown in Fig. 189.

Flat wicking is placed under bulkhead boundary angles the full width of the ship, and a short piece of ball wicking at all corners and staples. No wicking is required under stiffener clip angles, provided they are erected and riveted before the bulkhead is erected.

Flat wicking is placed the full length under the center line bulkhead boundary angles, the Fuel Oil Bulkhead boundary angles and under stiffener clip angles around Fuel Oil Bulkhead.

It is also used under Boiler Saddles and under the Engine and Thrust Foundations.

Where the clips to the Tank Top of the Auxiliary Foundations can be caulked all around, no stop-water is required. Where only the inside end of the clip cannot be caulked, a short piece should be placed crosswise under the clip.

Bulkheads below the Lower Deck are of three kinds, and are treated differently. When *water-tight under pressure,* stop-waters are used and the rivets in the bulkhead plates are countersunk. *Water-tight bulkheads not under pressure* require no stop-waters except at Tank Top and members adjoining bulkheads under pressure. The rivets in the top of the Shaft Alley are countersunk for another purpose. *Oil-tight bulkheads* require oil-stops and the rivets in the bulkhead plates are countersunk.

A short piece of flat wicking is placed in the seam between the shell plates opposite the boundary angles of the bulkheads under pressure. (Fig. 191.)

Flat wicking is placed under the outboard end of the compensating bracket clips and between bulkhead boundary angles and the shell (Fig. 191) on the caulking side of bulkheads under pressure.

CHAPTER XIII

Hold Stanchions; Foundations

In the Cargo Holds at the ends of the Cargo Hatches, it is customary to fit hold stanchions to support the framing in that vicinity. These stanchions are usually built up columns of plate and angle and are well secured at the head and foot. A wide bracket at the foot carried on to the doubling plate on the Tank Top carries the weight to this part where it is reinforced by extra heavy floor plates and sometimes additional stiffening by angle bars on the keelson. This carries the weight of the heavy deck beams, and framing around the hatches directly down to the bottom plating where the weight comes on the water.

Between the Upper and Middle (or second) Decks, stanchions are fitted on the center line of the ship, directly over the stanchions in the Cargo Hold. These are generally of the same type and the foot of the upper stanchion is secured to the head of the lower one by riveting through the deck plating.

Foundations might be classed as " main " and " auxiliary." The main foundations are those under the Boilers, Engine, Condenser, etc. Foundations under the Boilers are often called " Boiler Saddles " when they carry the round " Scotch Boiler " due to the round shape, but it is a term which is commonly applied to any Boiler Foundation. These Foundations are shaped to suit the Boiler according to the design used and are well braced to withstand the tendency to move when the ship is rolling and the Boiler is swinging with the motion of the ship. The Foundations must be very strongly braced athwartship as the greater part of the motion is in that direction.

Engine Foundations are more complicated when carrying a " reciprocating " engine than for a " turbine " because allowance has to be made for a passage of the crank on the engine and it means a weakening of the girder to allow for a deep cut in way of the crank. This is recompensed by additional stiffening. In vessels where the shaft is high from the floor, the Engine Foundation must be well braced both athwartship and fore and aft. With a turbine the center of weight is much lower than with a reciprocating engine, due to the height of the cylinders in a reciprocating engine being so far above the base plate of the engine.

As much of the weight in a reciprocating engine lies in the cylinders and valves, this matter of the height above the Foundation is a very important item to be considered when designing the Foundation to withstand the effort of the engine to break loose when the ship is rolling violently from side to side.

Main Foundations are made of heavy plate and angles. The angles are riveted with countersunk points through the flange which rests on the Tank Top, as that angle must be caulked to keep the Tank water-tight. The other riveting in the foundations is "snap." The plates are "lightened" by holes as much as possible, to save weight and for access when riveting and painting. The foundations are built to suit the general layout of the Engine and Boiler Rooms. The Walking Flat is laid afterward and brings the engineers and others at the proper height to do their work on the machinery.

The Foundation for the Main Condenser is heavily designed as this is one of the heavy items of the Engine Room installation.

The Foundation for the "Thrust Block" is strongly built in order to take the power exerted along the propeller shaft into the ship structure and force it ahead. All the power that is needed to drive the ship must be transmitted through this Foundation and for that reason it is one of the most important of the Foundations.

In the Shaft Alley "stools" are built; making a box-shaped Foundation of sufficient height to take the "steady bearing" of the Tunnel Shaft. The number of these stools varies according to the length of the "tunnel" shafting, but there is usually one on each piece of shafting, between the joints.

Auxiliary Foundations are those for the auxiliary machinery in the Engine Room such as the pumps, pipe manifolds, etc. These are either built up from the Tank Top or supported on the frames at the side of the ship, according to the location of the different machinery which they are to carry. These Foundations are often made up of a few angles riveted to the hull structure and drilled for holding "down bolts" for the machinery.

Foundations under Bitts and Chocks on the weather decks are only " doubling plates " riveted to the deck and carried far enough to run across nearby deck beams.

Foundations under the Windlass on the Forecastle Deck are braced under the deck beams and usually supported by stanchions or bulkheads to the deck below, thus tying a considerable portion of the ship's structure together to withstand the working of the Windlass. Cargo Deck Winches are held by local Foundations but are not supported by any special stanchion.

The Foundations for all this heavy machinery on the decks generally consists of a "doubling plate" on top of the deck and additional framing (as fore and afters) under the deck to catch the " holding down " bolts in the bed plates of the windlass, winches, etc.

Foundations for the Steering Gear at the after end of the ship vary according to the type of gear installed. This is also well secured to the ship structure as there is considerable strain due to the work done in moving the rudder when the vessel is at full speed.

CHAPTER XIV

Deck Beams and Plating

Deck beams are spaced the same as side frames, there usually being one beam on each frame. These beams are of angle bar or channel bar. On ordinary merchant vessels of to-day, channel bars are often used for the heavy decks and angle bars for the lighter decks above.

Beams are connected to frames by means of " beam brackets," or " knees," which are cut the shape of a triangle and fit into the corner formed by the junction of the frame and end of the beam. They are usually connected by snap rivets with an equal number in beam and frame.

Fig. 192 shows a beam bracket connection, the side frames running up to the deck above. The beam and frame "look" in opposite directions and the plate bracket is fitted against the back of each member.

The "shell angle," running intercostal between frames, is against the plate. The "deck stringer bar" is carried continuously along the inner flange of the frames, the intercostal angle being used to connect the side and deck plating and the continuous angle carrying the longitudinal strength.

In all such cases the deck stringer plate is cut in an oblong-shaped hole to allow passage for the frame. To fill the opening and to prevent a place for collection of dirt and moisture, hence corrosion, the space between the frames and the angle bars is filled with cement and sand.

A type of construction common on naval work but not often seen on merchant ships as it is more expensive, is illustrated in Fig. 193. The beam is split, bent to a radius and a piece welded in at the end. The angle shown here is

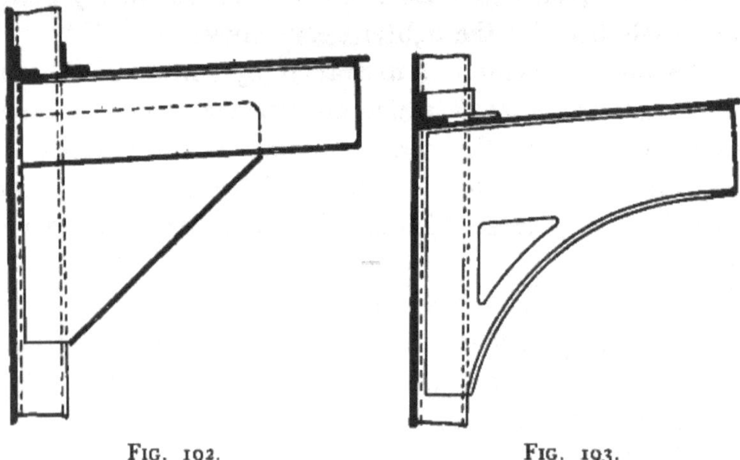

FIG. 192. FIG. 193.

"box-ended" against the frame. Where the construction calls for a continuous angle, the box-end is carried against the back of the angle, instead of fitting a short clip as shown here.

The "molded" line of the frame station is the governing location of the heel of the frame and the beam is offset from it by the thickness of the bracket.

Whole beams or "main beams" extend the width of the deck in one piece. "Half beams" are fitted in way of hatches and extend from the sides of the ship to the fore-and-aft girder of the hatch, where it is attached by means of double clips. If there is any opening or local construction which requires it, the beam is turned in the opposite direction from the others. In this case, it is called " reversed."

It is customary to put a " crown " or " camber " in all beams on the weather decks of about ¼ in. to the foot. This is in the form of a curve. Some ships have a straight pitch with ridge on center line, and again a few ships have no camber on weather decks. Beams on the Lower Deck are carried level, without " camber."

Deck plating is laid in the same style as described in shell plating; sometimes with a " raised" arrangement of strakes and in other cases " joggled " according to the machinery and custom in the yard in which the ship is being built. Seams in the plating are usually single riveted with a double-riveted butt lap.

Around hatch corners and under foundations for deck winches, bitts, anchor gear, etc., the plating is reinforced by a doubling plate, which is generally of the same thickness as the deck plates and extends for a few beam spaces beyond the point where the extra load or strain will come. It has been customary to cut the deck plating at hatches and other openings on a radius of about 6 in., but in some of the more recent boats, the opening is carried to a sharp corner.

The center-line plate is generally laid first and the other plating is laid in strakes from that plate working out toward the side of the ship. The strakes are lettered and then each strake is numbered according to the quantity of plates in that strake. The plate nearest the shell is called the "stringer" plate. In some designs, it has a greater thickness than some of the other plates, especially when it is on the weather deck and is in contact with the "sheer" strake of the shell plating. This combination of heavier plates at the corner of the box girder (as the ship is figured for strength) gives more material where the twisting and bending is most likely to occur.

The spacing of the riveting of the deck plating to the beams is the widest in the ship, being about 8 diameters of the rivet, which would mean 6 in. apart for a ¾-in. rivet. The riveting of the seams and butt laps have the regular water-tight spacing.

When the deck plating does not extend for the whole length, but the whole deck is covered with wood planking it is necessary to fit narrow plates (about 9 in. wide) around the ends of hatch coaming and other openings in the deck or at the end of the deck, if it is one of the raised decks, such as the Bridge or Boat Deck. This is done to allow space and material to secure the bolting for the ends of the deck planking.

Wood deck planking is held in place by "deck bolts." These are small, galvanized, with a square section under the head. Round holes are drilled through wood and steel, with a larger hole on top of the deck for the head of the bolt. The bolt is fitted with a grommet under the head, the square section of the shank preventing the bolt from turning. A grommet with washer and nut is fitted on the end. A wood plug is coated with white-lead and driven tightly into the hole in the deck and is planed off when the top of the wood deck is surfaced. The cracks between the planks are filled by "caulking" with cotton and oakum threads and then filling the upper part of the space with hot pitch or marine glue, which forms a watertight surface. The edges of the sides of the planking is often "outgauged" or beveled before it is laid, thus forming a space for the caulking. (Fig. 194.)

The steel weather decks, according to the design, are

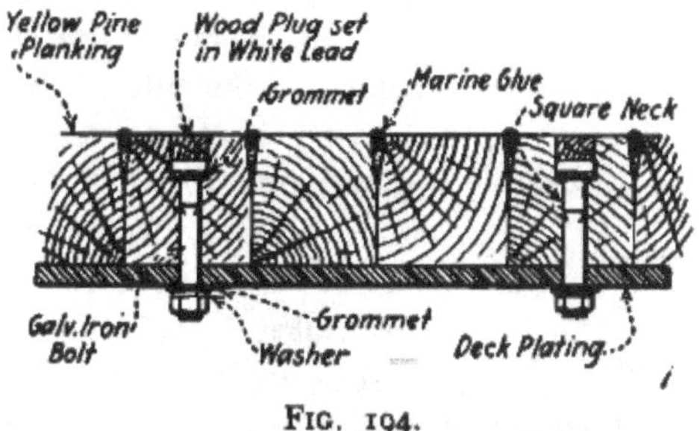

FIG. 194.

water-tight and are caulked the same as any bulkhead or shell plate.

Where the weather deck is covered with wood planking it is laid with a finish as shown (Fig. 195), the open space of bare plate being called a "water-way." The sheer strake of the shell plating is finished with a "half-round" bar for the sake of appearance. The stringer bar is against the shell plate and the "waterway" bar is used to hold the deck planking and act as a finish for it.

The arrangement and shift of butts in adjacent strakes being at least two frame spaces apart, is carried out for the same reason as described on the shell plating; that is, to avoid any weak part in the structure from having the breaks in the plating too close together.

As the drafting room generally orders these plates with only ½ in. to spare on the width of the plate and 1 in. in the length, it is necessary to be very careful when laying out the template on the plate so that there will be ample material to work on.

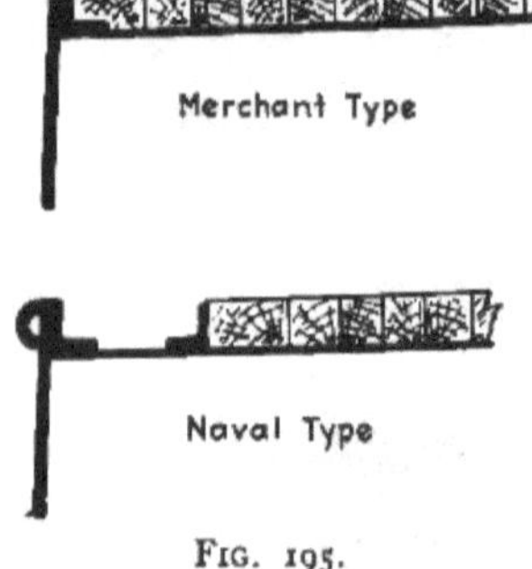

FIG. 195.

The Lower Deck is secured to the side of the ship by means of intercostal angles against the shell, between the heel of one frame and the toe of the next frame. These are called "shell chocks" or "shell angles." A "stringer bar" which runs continuously the length of the ship and catches one rivet at each frame, being riveted securely to the deck in the other flange gives additional strength. The space in between these angles and frames is filled with cement. This is done for drainage.

The weather deck is generally secured to the side by a continuous bar which fits against the shell plating as the frames stop below that deck except in way of the raised decks as the "Poop," "Bridge," and "Forecastle," where the same construction is carried on as for the lower deck.

The raised decks are generally of lighter construction than the main part of the ship, and the beams of these decks are ordinarily of regular angle bar. The plating is considerably lighter than that of the lower decks.

CHAPTER XV

Tank Testing

In order that the builders and inspectors of a ship may be sure that the workmanship is good and that all joints are tight, it is customary to test the various oil and water tanks in the Inner Bottom, the Fore Peak Tank, After Peak Tank, and all other parts of the ship that need to be tight against leakage. This is done while the ship is still on the building ways as the water used in making tests

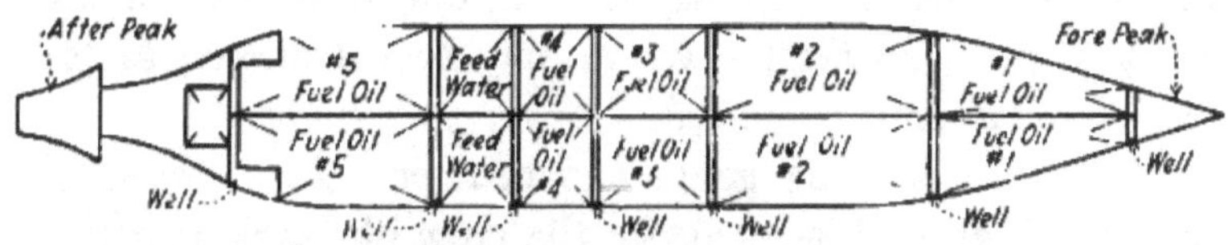

can be drawn off and need not be "pumped" out as would be necessary if the testing were done after the ship had taken the water. Also repairs are made more easily and satisfactorily before launching, especially the under water portion of the hull.

The work of the inspectors comes when the tank is full of water and ready for "final inspection." The work of the "tank testers" really comes before the water goes into the tank, in preparing the tank for the test. These men test, not by water, but with testing knives, along the joints and rivets.

For a man to be a thorough "tank tester" and of benefit to the ship building company, he must have a good knowledge of not only *what* he is doing, but *how* to do it correctly and *why* he is doing it. It is only by intelligent work that a man can be a thorough workman.

The compartments on *top* of the Tank Top in the ordinary merchant vessels are Cargo Holds, Engine and Boiler Rooms. The Cargo Holds are free of any special framing against the Tank Top except the frames at the side of the ship and the bottom end of the hole stanchions; but the Engine and Boiler Rooms are more difficult, as they have the foundations for the engines and boilers riveted to the Tank Top. The Engine Room may be considered as the most troublesome.

After the riveting gangs and caulkers have completed their work and all is ready for the crew of "tank testers," the Tank Top plating should be very carefully cleaned and swept free from all dirt, right down to the "bare" plates.

It is then the work of the crew to go along each seam in the plating, along the "toe" and "heel" and around the ends of every angle bar that comes into contact with the plating of the tank. All rivet points should be carefully inspected to be sure that they will be "tight."

This work is done with the "testing knife," a small, very thin-bladed knife which is of good steel and can be thrust into a very small crack or joint in the work. Where such openings exist they must be carefully caulked and left in good condition.

While a portion of the crew is working on the top of the tank, other men are inside the Inner Bottom going carefully over the watertight floors at the ends of the tank, working on the side of the floor which is outside the tank, and also again on the inside portion of the floor.

Other men are testing the work of the centerline keelson forming one side of the tank, and others are examining the outside of the shell plating.

(It should be remembered that after a tank has been tested and "passed," the centerline keelson need not be examined again before testing the tank on the opposite side; nor should the watertight floors need another examination before testing tanks which are forward or aft of the tank which has been tested. After these parts of the ship's structure have been found to be tight during one test, they may be considered as all ready for the next compartment.)

During this time the gang of "cleaners" (laborers) have been all through the compartment taking out all dirt and refuse which may have accumulated during the building of the ship. The tank should be perfectly clean inside and tight in all its joints on the outside before filling it with water for the "pressure" test.

After the tank has been cleaned and tested by use of the testing knives, it is ready for the water test.

It is the general practice to test first the tanks under the Boiler Room, thus allowing the boiler makers to get in there for work as soon as possible. Next come the tanks under the Engine Room, and then other tanks as they are ready.

The duration of the tests of the tanks will depend upon the word of the inspector. If the appearance of the tank is good, he may require only about one-half hour; if he is suspicious of the work, then he may need two hours, or more, before he will say "O. K."

Between the tanks in the Inner Bottom there is a "sump" tank, which is only one frame space in length and runs from the centerline keelson to the ship's side. These sump tanks are tested separately from the tanks on either side. The previous tests show the tightness of the watertight floors forming the sides of the sump tank, and when the pressure test is given to the sump it is only to test the tank top and bottom shell plating.

It is customary to leave one rivet hole open in the forward end of the tank and plug it with a soft wood plug. The inspector will go to this plug and pull it out. If the water spouts out with force then he has proved to himself that pressure is on the tank. If the pressure shows at the high end of the tank, then it must necessarily be on at the low end. One of the workmen will replace the plug in the hole when the inspector has seen the water and is satisfied, probably only a minute or so. After the test this plug will be removed and the hole riveted up.

It is also customary to fit "bleeder plugs" in the shell plating of the ship at the bottom of each tank, so that the tank may be easily drained after the test. These plugs are made about 2½ inches or 3 inches in diameter, according to the size of the tank, and are screwed into the shell plating of the ship. To give additional material for threading, a "doubling" plate of the same thickness as the shell plate and about 6 inches square, is riveted onto the inside of the shell. After the test these plugs are removed and the water is allowed to drain out onto the shipway. The plugs are then red-leaded and screwed in tight to remain there indefinitely, as long as the ship is afloat.

Stop-waters of lamp wick, soaked in linseed oil, are often used at odd corners and under watertight angles where it is necessary that the work shall be watertight. It is difficult to get into such places with tools (although the tools are used for caulking), and as a safeguard these stop-waters are also placed.

It is necessary that the tank tester should check up the work as he is inspecting it. However, it is the duty of the foreman riveter to be sure that his work is complete before his riveting gangs are through and taken off the job.

If the stop-waters are not in place, the rivets in the near-by locality must be taken out, the stop-waters placed and then other rivets driven to finish the work.

It is good practice to drive about six rivets in both flanges of the watertight bulkhead boundary bars up from the Tank Top plating to draw the bar into place and to help make the work tight. Watertight corners at these bulkheads are especially troublesome and should have close attention, otherwise the water will seep through back of the angle bars and will get into the tank next to the one which is being tested. Watch carefully the corners of all these large bulkheads, as time spent at these places will be well rewarded later on when water is in tank and it is found that no trouble is coming from the bulkheads.

The method of filling tanks varies in different shipyards. In some a wrought-iron pipe is threaded and fitted to the bleeder plug, with a by-pass for drainage afterward, and the tank is filled from this pipe. In other yards the tank is filled through a hose down into an open manhole, the bleeder plugs all being in place.

A hole is drilled in some part of the tank top plating, near the high end, and a wrought-iron pipe about 1 inch to 2 inches in diamater is threaded and fitted to it. This pipe is carried up and a large funnel, about 18 inches in diameter, is secured to the upper end.

During the water test some yards make a practice of using a certain manhole cover with a 2-inch hole threaded for the standpipe. This cover is moved from tank to tank as each is being tested, avoiding the necessity for drilling a special hole in the Tank Top plating, the regular cover for that manhole being put in place after the test.

It is necessary to fill this pipe until the water shows in the funnel, the testers and inspectors then knowing that the tank is full and under pressure. When this pipe is full, the surface of the tank has a pressure of about one-half pound per square inch for every foot of height of the water in the pipe.

The height of the standpipe with the funnel is determined by the requirements of the Bureau under which the ship is being built.

For example, the following are the general rules used by one of the marine classification societies:

Double-bottom Tanks Carrying Water. To be tested with a head of water to the height of the load water line.

Double-bottom Tanks Carrying Fuel Oil and Deep Tanks Carrying Fuel Oil. To be tested with water by a head of water extending to the highest point of the filling pipes, twelve (12) feet above the load line or twelve (12) feet above the highest point of the compartment, whichever of these is the greater.

Deep Tanks and Peak Tanks Carrying Water. To be tested by a head of water eight (8) feet above the top of the tank, but not less, in any case, than to the height of the load water line.

Bulkheads, Shaft Alley and Weather Decks are to be submitted to a "hose test" when water from a fire hose is turned on the plating and aimed at the joints. At this time an inspector stands on the other side to watch for any leaks through the joints.

The inspectors are often from the prospective owners of the boat and also from the marine classification society which inspects the vessel while it is being built. These marine societies are "The American Bureau of Shipping" and "Lloyds Register of Shipping," a British Corporation under which many vessels are being built.

These inspectors will also be "on the job" with their test knives, and any leak which they may discover must be at once remedied.

Before the time for filling the tank with water be sure that all manhole covers are in place. If they have not all arrived, then fit a temporary cover over the manhole, as the tank must be closed on all sides, top and bottom.

Manhole covers are sealed watertight by using a "grommet" of pasteboard and red lead, or lampwick and red lead, which is placed under the cover, around the edge, so that no water can seep through between the cover and the Tank Top plating.

After a final inspection, the water is poured into the tank. If an open manhole is used with a hose into it, the water is allowed to come up to the top of the tank and then the hose is withdrawn and the cover put on, the balance of the filling being done down through the funnel at the top of the stand-pipe. When this method of filling the tank is practiced not only is pressure insured when the tank is full, but an air vent also is provided, and for this reason the pipe should be placed at a high portion of the tank (usually at the forward end, as the vessel is built at an incline toward the stern).

Soft pine-wood plugs about $\frac{5}{8}$ inch in diameter at the small end are kept nearby so that in case any rivet hole has been overlooked the plug can be driven in place, the soft wood swelling so that it will hold against the pressure tending to drive it out. Sometimes the high head of a bad rivet will have been chipped off and then the workman has neglected to "back-it-out" ready for driving another rivet in its place. The water pressure will work the rivet loose and the shank and point will fly out, leaving the hole open. This may also be closed by use of a wood plug.

Candles and electric flashlights are used by the tank men when working and inspecting in dark places and through the Inner Bottom if they are without a "portable" electric light on a wire.

After the tank has been filled comes the final test of the workmanship, and to the tank tester who has put in many weary hours on the job the result of his conscientious labor is apparent.

While the tank is filling sometimes small, unimportant leaks will develop along the edges of the plating and at the points of some of the rivets. It is the duty of the men in the gang to stop up these leaks by caulking, generally using hand tools instead of air hammer, as light blows are needed at this stage of the work, rather than too severe a blow on the metal, as this would tend to open up a more extensive leak.

All men who are good tank testers should have had experience in chipping and caulking with the air hammer; but for most of the tank testing it is better to use the hand hammer, as the latter tool is more easily controlled and with it it is possible to give a light blow when a heavy one would do more harm than good.

When using the hand hammer grasp it well back on the handle so that a good swing can be given, and yet have the hammer head respond well to guidance. A little practice will give a sense of touch that is necessary for good work.

When caulking rivet points use a "frenchman" which has a sharp edge and is rounded, instead of being in a straight line at the point, so that it will tend to follow the edge of the rivet.

The caulking tool should be held lightly in the other hand, the back of the hand being downward, the tool passing between the thumb and the first finger, passing over the last joint of the second finger. The third and fourth fingers being closed and held lightly against the palm of the hand.

By holding the tool in this manner it is possible to obtain a much better view of the work and if the hammer should miss the head of the tool, a blow on the part of the hand as exposed in this way will not be as serious as when received across the back of the hand.

It is best to always wear eye goggles when on this work, as many times small chips are liable to fly up in the face of the workman. In the line of "Safety First" the worker should always be sure that no other man is working directly over him or that no loose tools are in a position where a slight jar will cause them to fall on him. This is always possible, as most of this tank testing is done near the bottom of the ship and there are often many gangs of riveters and others working on different parts of the ship above.

Clean the metal around the rivet as well as possible with a rag of cloth or a piece of cotton waste so that it is all clear; then pass around the circumference of the rivet point, striking lightly. This will outline the edge of the metal of the rivet. After passing around once then follow it with a trip around, giving heavier blows to force the metal, taking care at all times not to cut into the edge of the plate but to work on the metal of the rivet. In this way a sharply defined line will show in the form of a ring around the rivet, the surface of the plate remaining smooth and clear.

As the rivet point has been driven into a countersunk opening, this driving of the material in the rivet down into the plate will force it deeper into the beveled opening, or countersink; and the farther down it is driven, the tighter it will become. Thus the "caulking" is accomplished.

A leaky rivet is possible only when the body of the hot rivet has not been driven in hard enough to swell out and fill the straight part of the hole; i. e., the body of the rivet is smaller than the opening. Water will leak under the head, along the body of the rivet and will appear as a leak at the point. If the plates which are connected by this rivet are so tight together that the water cannot run along the body of the rivet and then between the plates, to appear out from between them, in the seam at some distant point,—then the leak can be stopped by caulking the point of the rivet.

Caulking the point forces the material in tighter against the steel plate so that there is no passage between the metal of the rivet and that of the plate.

Many times a rivet head is not correctly formed in the rivet-making machine, and the resultant head is beveled outward toward the edge. When this happens and the rivet is not discarded by the riveting gang, the rivet will be driven up hard, the point properly hammered, and yet the rivet is very liable to leak and cause trouble as water can get in under the head. The under side of the head of the rivet will touch the plate and bear hard on it, but a crack will show around the outer edge.

When this crack is less than $\frac{1}{16}$ inch it can be caulked; but when it is greater it is difficult to obtain a good job, and it is best to cut out the rivet and drive a new one.

When caulking the head of the rivet use a blunt, smooth tool so it will not mar the surface. Do not use a "fuller," as that tool will form a ridge on the head and will often cause it to be condemned by an inspector, due entirely to the appearance.

When caulking the rivet head work around the side, taking care to keep some distance up from the edge, and working a little of the material of the head down toward the plate, thus filling in the opening between the under side of the head and the plate. This operation should be continued until the material is down hard against the plate and the crack has been entirely filled up. This is known as caulking "highhead" rivets (usually "panhead") and can be done to pass inspection if the workman is careful not to scar the material, as often a good, watertight job will be thrown out if the inspector does not like the appearance of the head.

(This kind of caulking is generally done before the water is let into the tank, as the rivet heads are usually on the inside of the tank.)

Hand-caulking a seam is similar to regular caulking as done when the caulker follows the riveting gang, except that when there is water in the tank a different condition exists, and also because the metal of the edge of the plate has been once worked over.

When a leak is found along a seam in the plating, clean the metal with rag or waste, then use a "fine fuller" or "fine splitting tool." As the line between the plates is straight, more material must be used to fill in the opening. (This cannot be caulked like a rivet point where the bevel opening helps to make a tight joint.) This additional material is obtained by roughing up the steel close to the crevice between the two plates, and then hammering this burr down into the opening, thus closing up the leak.

The size of the burr formed for material for caulking is governed by using about one-half the width of the part already caulked (about $\frac{1}{8}$ in.), which would give a burr of about $\frac{1}{16}$ in., this being sufficient material to stop up a small leak.

"Pin Leaks" are often formed where there is an imperfection in a small part of the steel plate, through which a tiny stream of water may squirt out. It is often possible to stop these by "scratch caulking." This is done by holding a chisel between the thumb and second finger, with the first finger on the butt of the chisel, and tapping the plate near the leak in order to rough it up a little and form a burr. This is then forced into the leak by a side sweep of the chisel, with the point passing across the opening of the leak, thus bending the burr over and into the opening. Little pressure is required to force the burr down into the leak, as the blow needed to bend it over will force it into the opening.

Sometimes an imperfection in the steel will cause a line of rough, coarse-fiber material, and when this happens to come on the caulking edge of a plate a leak will develop which the caulker who first went over the seam would not have expected to find there. When the caulker goes over this edge for the first time and meets this rift in the material of the plate he does not notice it; but the coarse fiber will bend back until it is forced onto the hard, firm material of the good quality of the steel. In this way a crack is often liable to form through which the water will be able to work its way when it comes time for the test.

The tank tester must lightly chip off this feather edge, back to the solid material, and then caulk that as though it were being done for the first time.

Care should be taken at all times to strike light blows on all parts of the work, as much damage can be done toward opening up a leak in another nearby part if the plates are too much jarred by the vibrations set up in the material by hammering on the surface.

It often happens that a leak will occur around a rivet

and the water will run along back of an angle bar and the tester will not be able to locate the place at which the water is coming through. Such leaks occur at foundation angles on the Tank Top, watertight floors, angle bars on the Bilge Keel, watertight bulkhead boundary bars, and in some cases where the "three-ply" riveting is bad (meaning where three plates or two plates and an angle flange are secured by through rivets).

Such leaks can be stopped only by the use of the "putty-gun." This consists of a small cylinder with a piston which is moved by means of a screw piston-rod operated by a crank handle. The forward end of the "gun" has a cap which may be screwed on and off, and in the center of the cap a $\frac{3}{8}$-inch pipe is carried out for a few inches, the outer end being threaded and slightly bent.

At the point where it is decided to try to stop the leak, a $\frac{3}{8}$-inch or a $\frac{5}{8}$-inch hole is drilled and tapped (the hole is drilled $\frac{1}{16}$ inch less). To operate the gun, the piston is run back, the forward cap taken off, and the "putty" put in until the gun is full. The forward end is then put on. The pipe nozzle on the end of the cap is now screwed into the hole in the material where the leak is to be stopped, and the piston is then forced forward by means of the screw, thus pushing the putty out of the end, through the pipe, and into the space where the leaking water is traveling along. This space, back of the angle or between the plates, as the case may be, is filled tight with the "putty," a second or third charge of the gun being used if it is found necessary.

The gun is then unscrewed and the tapped hole is filled with a metal plug which is screwed into it.

"Putty" is a stiff mixture of red-lead powder and shellac, thoroughly mixed before placing in the gun.

Sufficient linseed oil is added to the powder to form a paste, and then the shellac is added in quantity enough to make the paste liquid so it will flow through the gun and the pipe down into the space and fill up each corner of the crevice. The red lead forms a watertight obstacle to the leak, and the shellac acts as a hardener, solidifying the mass after it has set for a few minutes.

During the pressure test, when any leaks are liable to show, it is the usual practice to have a few small bags of sawdust handy to spread on the Tank Top if it is desired to dry up quite a large-sized pool of water on the plating.

When there is an excessive number of leaks and the general appearance of the tank is unsatisfactory to the inspectors, then the water may have to be run off and some of the rivets cut out and replaced. The tank is then refilled and again tested for watertightness.

Careful preliminary work shows during the pressure test and the tank-testing crew must be composed of skilled men and conscientious workers, as much of their work is done "on honor," individually and collectively, and the reuptation of the firm is often in their hands. They are the "working inspectors" for the shipbuilding company.

CHAPTER XVI

Launching

As explained in Chapter IV the two common ways to launch a ship are the "end on" and the "side" method.

When a " side launch " is necessary the ship is built on an "even keel," but for an "end launch" the ship is built with the keel on an incline of $\frac{1}{2}$ in. to $\frac{3}{4}$ in. per foot, so as to be in the correct position for launching. It would be impossible to raise the ship to this incline after it is built, and unless thus inclined the ship would not tend to move down the ways of its own weight.

After the hull has been thoroughly inspected and "passed," the outside plating is painted and the ship carpenters fit their launching ways which are made up of two parts; one called "standing" or "ground" way; the other the "sliding" way or "cradle."

The ground ways are made up of heavy plank built up to the necessary thickness and firmly bolted together to give a section of about 12 in. thick and about 36 in. wide. On top of this a sliding surface of oak plank about 3 in. thick is laid. On the outside edge a heavy plank is bolted, projecting above the top about 2 in. This is to act as a guard to prevent the sliding ways from running off the ground ways. These ways are built in sections about 35 ft. long and are connected by means of heavy ugs, bolted to the way and bent out to bolt to ljoining lug on the next length of way. (Fig. 196.) ie outboard (or water end) sections of the ground are secured in place and the other sections are held it them, fore-and-aft. To steady the ground ways spreading, shores are fitted at each bent of the ship-

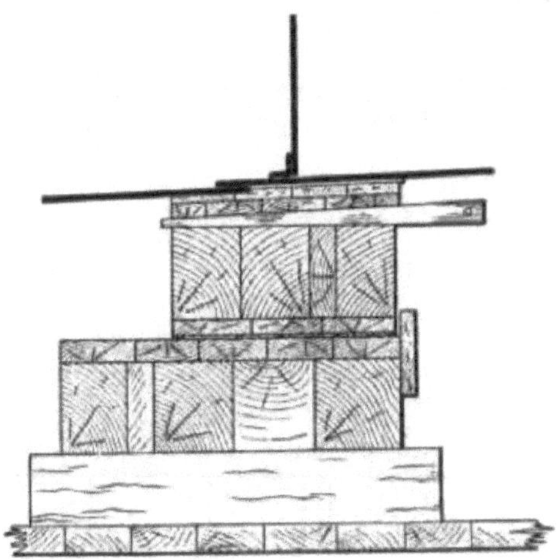

Fig. 196.

ie outboard, or water end of the ground ways is set ce below the high-water mark by building a coffer-around the piling, pumping out the water, sawing : piling to the grade set by the engineers, spiking the bents to pile heads, then laying the ground ways olting them in place. As the outer end of the ground must extend sufficiently far to allow the ship to be borne as she leaves the ways, they must be carried me three or four lengths. (Fig. 197.)

The ground ways are spaced about 20 ft. apart, on the outside, the distance varying with the size and beam of the ship to be launched. In some yards, the ground ways are built with a "camber" or slight rise in the

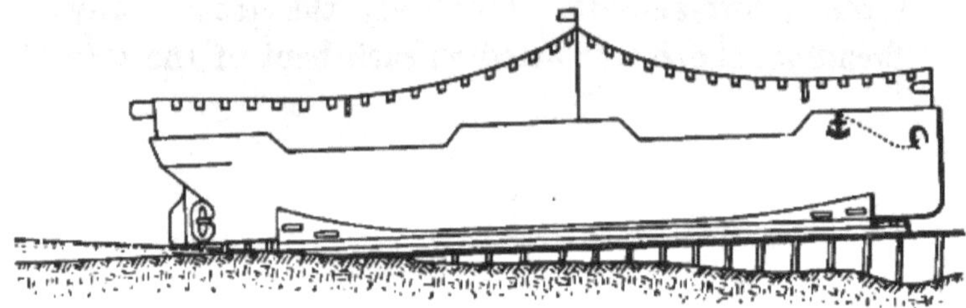

Fig. 197.

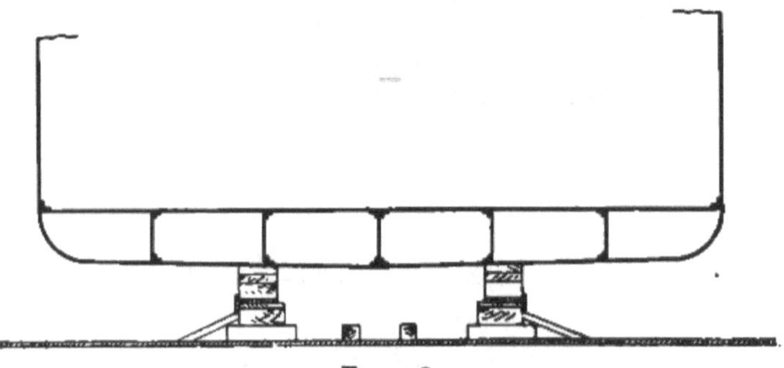

Fig. 198.

mid length, to offset a tendency to "sag" when the weight of the ship is on them, and they tilt toward each other about ¼ in. (Fig. 198.)

The "sliding" ways are built up in the same manner with an oak sliding surface but without any guard. They are made in sections for easy handling. After the ground ways are in place under the ship the sliding ways are hauled up on them from the water end by means of a rope and a power drum.

Between the sliding ways and the bottom of the hull of the ship "packing" which is made up of planks about 2 in. thick by 9 in. wide is edge-bolted together for a width equal to the sliding ways. This packing is fitted to the shape of the plating, allowing for plate landings and butts, so there is an even and firm contact the full length

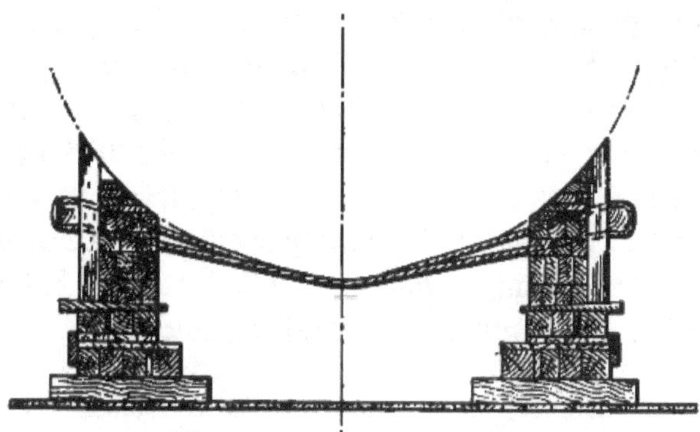

Fig. 199.

of the way. Amidships where the ship is "full" the packing is only about the thickness of the planks, but towards the ends of the ship, where it changes shape, more material is fitted in.

The cradle usually extends near the bow and stern and here the packing is so steep that heavy timbers are carried vertical and a horizontal piece is carried across outside of them. A heavy rope or chain is wrapped around the horizontal piece, down under the keel and up to the packing on the other side. This packing at the ends is called "Poppet." Those at the forward end, port and starboard, are called "Fore Poppets," and those at the stern are called "After Poppets."

The rope, or chain, is necessary, as the steep sides of the ship at the ends would force the cradle off to the side if it were not held in place. The downward pressure on the chain draws the poppets in toward the hull and holds them in place. (Fig. 199.)

In order to make the ways slide easily it is necessary to have plenty of grease on them. This grease, with a body of heavy tallow, is floated on the top of the ground ways, about $\frac{3}{4}$ in. thick with a float and the "spacer bars" of iron, $\frac{3}{4}$ in. square and of sufficient length to stretch across the ground way are placed at intervals to take the weight of the sliding ways until time for the launch. Launching grease is applied to the bottom of the sliding ways and they are laid in position.

LUBRICANTS ON SHIPWAYS FOR "END" LAUNCH

Lubricant	How Applied	Quantity and Position
1. Stearine.......	(hot, with brush)	Two coats about $\frac{1}{16}$ in. thick on ground ways, from midships aft.
2. Stearine.......	(hot, with brush)	Two coats about $\frac{1}{16}$ in. thick on bottom of sliding ways.
3. 1 part lard oil... 5 parts tallow.	(Floated on hot)	About $\frac{5}{16}$ in. thick on ground ways, for full length of sliding ways.
4. 1 part lard oil... 3 parts tallow..	(Floated on hot)	About $\frac{5}{16}$ in. thick on top of ground ways, on top of No. 3, for full length of sliding ways.
5. Neptune Grease.	(Spread on)	About $\frac{1}{8}$ in. thick on sliding ways only.
6. 1 part lard oil.. 5 parts tallow..	(Floated on hot)	About $\frac{3}{8}$ in. thick on outboard end of ground ways.

Between the packing and the sliding ways long wedges are driven in at intervals of about 7 ft. These are bored with a hole through the head and a rope is passed the length of the ship, in order to be able to recover them after the ship has been launched. These wedges are counted off in groups of about fourteen in each and the heads of the wedges in each group are painted red or white, in alternate groups, so that when the men are ready to drive the wedges up to take the weight of the ship, they will know what wedges they are to work on.

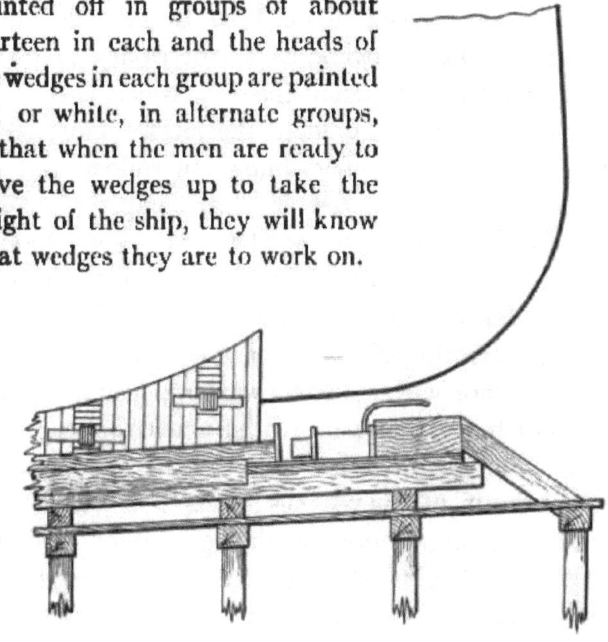

FIG. 200.

There are a number of different methods for securing and releasing the launching ways, in different shipyards of other countries, but in the United States it is a common practice to bolt a heavy oak plank, about 5 in. thick by 36 in. wide to the forward end of the under side of the sliding ways, and to the top side of the ground ways, leaving the top of the plank clear from the bolts and end of the sliding ways enough to be able to saw it across. There are two such planks, one on each side of the ship. The line for sawing is generally marked down the sides and ½ in. intervals marked off on it, so the sawyers will be able to keep even when sawing, as it is necessary that both planks shall be sawed through at the same time. (Fig. 200.)

Before the day of the launch, the marine engineering staff have fitted the propeller, hung the rudder and the ship carpenters have secured it "amidships" (in line with the keel) with two heavy wood battens. The "bleeder plugs" have all been replaced in the hull since draining off the water for testing the Inner Bottom. The necessary bitts and mooring chocks have been fitted on board and a kedge anchor with sufficient wire rope cable is slung along the bow to use in case of need when the ship is in midstream. The painters have been over every square foot of the bottom where possible to reach (as the shores are taken down, just before the ship goes overboard, the painters are just behind the workmen and paint all the places where the ends of the shores were touching and could not be painted before). The ship carpenters have removed all the staging around the stern and drawn away from the ship's side the cross-spalls and piled the staging plank on them, one wide and three high. As the side staging is parallel with the ship, it is not necessary to remove any of that. All compressed-air hose is taken off the ship, back to the stand pipes, and all ladders or anything else which touches the ship has been removed.

During the low tide just preceding the launch the tops of the ground ways which have been under water and therefore not greased, are heated with flat coke ovens to dry them off and then the grease is put on to match with that already on the ways further up from the shore. To prevent a possible "sticking" on the ways, two hydraulic jacks have been installed, one on each way, at the head, with the plungers near the end of the sliding ways. These jacks are connected up to a pump which can deliver a kick of about 200 tons for each ram. These jacks are now tested to be sure they are all ready for service, if called on.

In order to prevent sliding of the cradle beforehand, "dog shores" are fitted near the after end of the ground and sliding ways. These are diagonal braces with ends resting against lugs on the two ways. The pressure is supposed to hold them in place, but to be sure they do not drop before the time, they are supported by a small prop.

Because of the extra, local weight on the stern, due to the stern frame casting, propeller, rudder and overhang of the stern the after end of the ship must be supported for as long a time and as much as possible.

To do this "tumble shores" are fitted. They are heavy, square timbers placed close together extending from the bed of the shipway to under the keel. They remain in position after the ship has been "wedged up" and keel blocks removed; and fall over backwards as the ship slides down the ways on its journey into the water.

To allow for easy movement the edge on which the

shore will turn is rounded over from the center of the end of the shore, thus giving sufficient bearing surface but no sharp edge which will tend to cause the keel to raise by the shore pivoting as it falls.

About two hours before time for the actual launch (this time varies in different yards according to the skill of the workmen), work is started on the cradle. "**Remove Spacer Bars**" is the first order given, when all the iron bars are drawn out from between the ways, laid off at one side and counted, to be sure they are all out.

"**Dog Shore Guards Stand By**" is the next order, when the men who have been detailed to remove the dog shores are assembled and their bosses make sure that all men are accounted for.

"**First Rally on the Wedges,**" follows, when the workmen who have been previously divided into gangs of about eight men on each ram (rams are heavy oak pieces about 8 in. square and 10 ft. long, with rope handles at the side for handling) start to drive in the wedges between the cradle and the packing. A rest period for the men is followed by,

"**Second Rally on the Wedges,**" when they are again driven in and any of them which had not been driven in straight the first time have been seen by the inspectors while the men were resting and the wedges are now straightened and driven in still further. A rest period for the men again follows. Another inspection.

(Some yards have a "Third Rally" on the wedges, but other yards prefer only two rallies.)

"**Remove Joggle Wedges and Grease Covering Strips,**"

is now heard and small wedges which have held the cradle centered right are taken out and the surface of the ways greased.

"**Remove After Shores.**" This means that the shores supporting the ship by bracing against the hull, at various places are taken down, starting from the stern, for about 40 ft. forward. The men then wait for another order.

"**Remove Shores and Timber Keel Blocks.**" The remaining shores are now taken out, working forward and keeping an advance of about the distance gained on the last order (40 ft.) ahead of the men who are removing the keel blocks, starting from the stern and taking out alternate blocks.

The Tide Report is now given to the man in direct charge of the Launch, so he will know the exact condition of the water over the ends of the ground ways.

"**Stand By**" is now heard. (This is about one-half hour before the plunge.)

Some yards use cast-iron boxes with an open top and a large screw plug at one end, called "Sand Boxes." These are placed on alternate keel blocks and on the side cribs which have been built up under the bilge of the ship, four on each side. If this yard uses these boxes, the next order will be,

"**Remove Sand Box Plugs,**" when the plugs are unscrewed. A report is made that all plugs have been taken out.

"**Spoon Sand and Remove Blocking**" is the next order, when the sand is spooned out from the box and the remaining keel blocks and the side cribs are taken down and removed out of the way of the hull when it is ready to launch.

If this yard does not have sand boxes the order would have been,

"**Remove all Keel Blocks and Cribs,**" when the only things now holding the ship from sliding are the saw-off planks bolted to the ways and the two dog shores braced against the two ways, on the outside edges.

"**Remove Dog Shore Side Braces Only**" is now heard, when the Dog Shore Guard (who are old and experienced hands at the business) carefully removes the props under the dog shores, the increasing pressure holding the dogs in place.

A final inspection is now made by the men in charge, one going along each side of the ship, and one going along the center line to be sure that all is clear.

"**Drop Dog Shores**" is then ordered, and the men detailed to look after the shores knock them down. If any of them stick, the men cut the shore with an ax until it falls down out of place. Two minutes is all that is allowed for this operation. Now come the final words:

"**Saw-off,**" and the sawyers at the saw-off planks at the head of the ways start to saw as fast as they can move, as the ship is now in a dangerous condition if anything holds her unevenly on the ways. These two cross-cut saws have two men at ends of the saw (some yards have extra handles with four men at ends of the saw). A man in charge of the sawing stands between the two gangs on the two sides of the ship and counts so they will saw in unison, keeping his eyes on the advance of the saw down the line of ½-in. intervals which had been marked on the side of the plank. Just as the saws reach nearly to the bottom of the plank a sound of cracking and tearing of wood is heard and the ship starts very slowly, gathering momentum as she proceeds until at the time when the bow reaches the water the ship is traveling fast. A 400-ft. ship will travel down the ways in about forty to fifty seconds.

Just as the ship starts on its first journey, the sponsor in the Launching Party does her share of the work and, leaning over toward the stem, hits it a sharp blow with the christening bottle, naming the ship as she does so.

In planning a successful launching it is necessary to forestall the likelihood of two possible accidents, capsizing and tipping. There is a distance during the travel of the ship down the way when the after end of the vessel is water-borne and the forward end is land-borne on the forward end of the cradles (called "poppets"). If the ship stopped moving anywhere during the time required to cover this distance, it would capsize, as it is supported only at both ends and not in the middle.

It is necessary to have sufficient depth of water over the end of the ground ways so that the after end of the ship will be water-borne before the center of gravity of the ship passes beyond the end of the support of the ground ways; thus avoiding a "tipping." If the vessel, "tips," the stern dropping down until water-borne, the bow rises and the position of the forward "poppet" as held by the rope cable, as before described, will tend to sag and drop. Then later when the after end of the ship is water-borne, and the bow of the ship comes back to its original position in the cradle, there is trouble if the cradle is not there to receive it. (Fig. 201.)

Neither of these accidents is likely to happen in a well-regulated launch, because the speed of the ship being fast and continuous, carries it through the danger space before it can start to capsize, and the drafting-room force having previously calculated a curve showing a tendency of a tipping moment and the condition which would cause this has been carefully looked into and avoided.

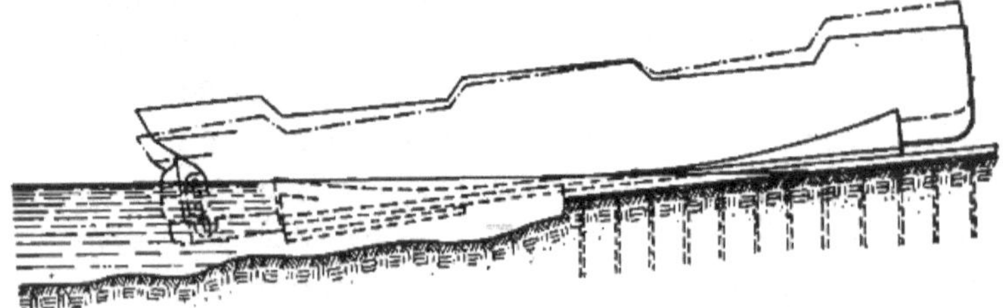

FIG. 201.

In yards where a broad surface of water is available for launching into, the ships are allowed to run free until their momentum has died away, and they may either drop an anchor or are towed back to the wharf at the yard.

Some yards, due to the narrow space in which they are able to launch their vessels, are obliged to use different methods for checking the momentum of the ship and it is customary to either secure cables to the ship and land which will retard the progress of the boat by breaking of the cable, or to provide bundles of heavy chain which are connected to the sides of the boat by a chain pennant, causing the bundles of chains to pick up, one after another and drag along the ground; care being taken to have a bunch of chains on each side taken up at the same time, in order to have the strain equal. (The writer has seen vessels launched on the river Clyde which were laid down diagonally to the shore line in order to give more distance, and these vessels when launched thus with bunches of chains would lose their momentum just after having gotten into the water.)

Because of magnetic influence it is customary to head a ship in the opposite direction after the launch and she is usually tied up at the wharf in this way.

In those shipyards where side launching is necessary, as in many of the yards on the Great Lakes of the United States, the ships are laid down with the keel parallel to the water front, allowing plenty of room between the side of the ship and the water for working and for handling the material.

As the vessels will be moved sideways, the keel is laid on a level line fore and aft, at a height sufficient to allow room for men to work under it, and under the inboard bilge (away from the water side), if the ground slopes toward the water.

One advantage obtained when building ships in this manner lies in the fact that it is possible to "plumb" bulkheads and other parts of the ship without allowing for the inclination of the keel as is necessary when building for an end launch.

When the time approaches for the launch, standing ways are laid under the ship about ten feet apart with a declivity of from $1\frac{1}{4}$ in. to $1\frac{1}{2}$ in. per foot, depending on weight of the vessel. They extend from the side of the ship, furthest from the water, to the edge of the shore. Due to the rise of the ways it is possible to obtain a height of about four to five feet under the bottom of the ship which, is good working headroom while the vessel is under construction.

The ways are placed so that they pass in between the keel blocks. The launching ways in this case are simply smooth platforms of wood which are placed under each side of the ship, near the bilge and the ship carpenters then fill in the intervening space between the ways and the bottom of the ship with the launching "packing," in the same manner that it is used for launching "end-on."

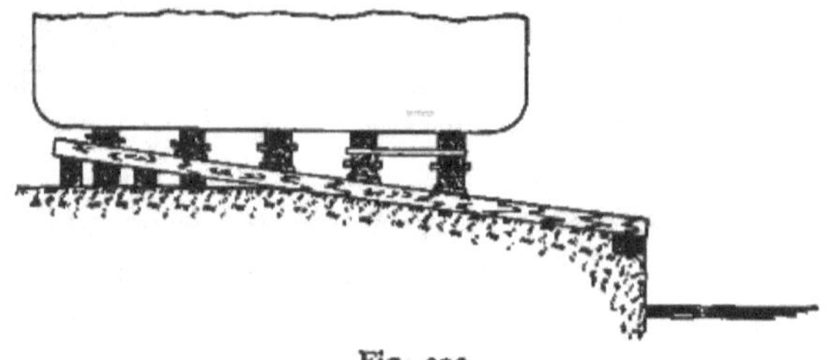

Fig. 202

Fig. 202 shows the blocking and ways for a ship which is to be launched sideways. The diagram (shown in Fig. 203) gives the arrangement for the launching release. The ground ways are shown at *AA*, the number varying according to the length of the ship; *B* is the line of the keel, *C* the strut placed against the keel and holding it from moving, the other end being held by the trigger *E* which is held in place by the heavy stake *D* driven into the ground and a manila rope doubled enough to give ample strength shown at *F*. This is carried back and secured to a stake *G* in the ground, located beyond the ship so that when he ship starts to move the workmen will not be in the path. 7 is a heavy wood plank which is placed close to the rope nd supported securely in place so that a workman with a eavy sharp axe can cut the rope clear through with ne blow, thus releasing the trigger, strut and ship.

In the case of a large, heavy ship there are three or four f these triggers, but for smaller ships there are only two, ne near each end of the ship.

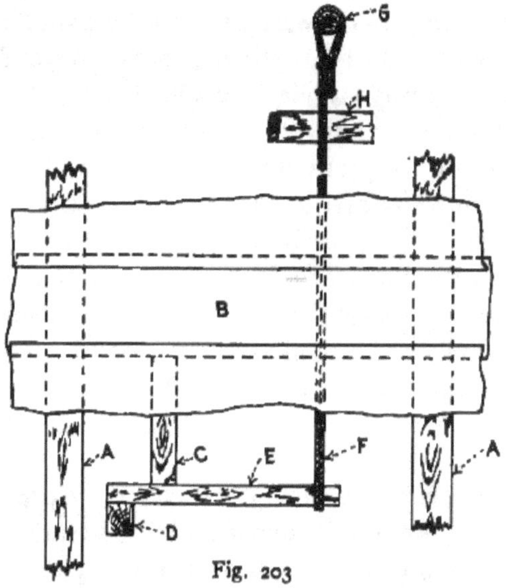

Fig. 203

The time from starting until the ship is in the water 'aries according to the size, but it is about ten seconds. ;ome vessels of smaller size (say 3500 tons) make the trip n six and a half seconds whereas other vessels of a heavier veight require more time as their motion is slow at the start.

The shore end of the standing ways should be as near he water as possible, one or two feet being a good height, but there have been launches on the Great Lakes where the drop from the ways to the water was nearly fourteen feet. This is not desirable and they were not large vessels that took the plunge. If the ship does not slip evenly, so that one end precedes the other, there is apt to be a considerable wrench to the whole structure and this is one of the features about side launching which must be carefully avoided.

As the center of gravity of the ship passes beyond the end of the standing ways, while they still support the land side of the ship, the hull tips sideways toward the water and strikes the surface partly on her side and bottom, the bilge entering the water first of all, Fig. 204. This is better than to have the flat, or nearly flat, surface of the bottom strike evenly on the water, but the momentum of the hull must be sufficient to carry it through the water far enough so that on the following roll of the ship, toward the land, the side of the ship will not be damaged by striking against the ends of the standing ways, which terminate on the land, at the water's edge.

As mentioned before, the end-on launchings are preferred where possible, as the ship is supported by the cradle until it is water borne and thus any chance of straining the structure as might happen during the plunge of a side launch is avoided.

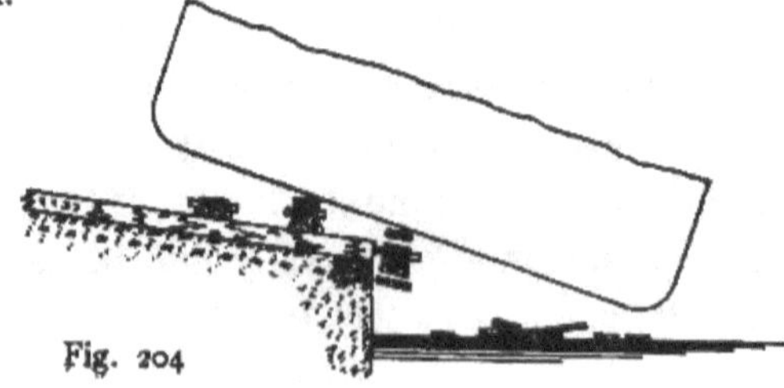

Fig. 204

CHAPTER XVII

Ship Fittings and Joiner Work

Ship fittings and shipfitting are entirely different. Ship fittings are the small articles which are fastened to different parts of the hull structure after the hull proper has been built and launched. These are in the way of equipment, both inside and outside the ship.

Shipfitting is the placing and fitting (making them fit when they do not at first fit) the small parts of the hull structure, such as angle clips, brackets, foundations, boundary bars on bulkheads, etc. Any of the work outside of "regulating" the plating after it has been erected, is "shipfitting." The process of placing ship fittings is also shipfitting work.

Boat Davits, according to a number of designs, are usually made at the Blacksmith Shop with all the small pad eyes and other things for the use of the rigger.

Rail Stanchions are made according to the type of rail, whether of rope, pipe or top rail of wood. These are to be drilled and riveted or bolted (if fixed or portable) to the sheer strake or coaming, according to location.

Awning Stanchions, with ridges and guy ropes are usually of pipe with a fitting at head and foot, so thay may be easily "struck down."

Bitts and Chocks are riveted to their foundations,

being set in a packing of red lead and canvas between the base and the deck plating.

Iron ladders for access to various parts of the ship, gratings in Fidley Hatch, skylights over the Engine Hatch, cargo hatch battens, fittings and other small steel parts both on the outside and inside of the ship are the work of the ship fitters.

Jack and Ensign Staffs are fitted to castings on the deck or into pipe supports. The location of these flags was changed somewhat since the necessity for bow and stern guns on the Forecastle and Poop Decks was paramount during the World War.

Side Accommodation Ladders have davits and fittings for use over the side and other chocks for stowage in on deck when not in use.

Much of the work just noted is not done by shipfitters, but by ship carpenters who make a specialty of this kind of work, and are a different gang from the men who work on the shipways, on the staging and similar work.

The work on the masts and their cargo booms with all the rigging, is done by the Riggers. This gang also reeve all the falls for the boat davits, flag halyards, rope rails, accommodation ladder and all other rope work about the ship.

So many types of ship fittings are required to suit local conditions that they are usually designed to suit the type of boat being built and detailed descriptions cannot be given. There is no set rule for ship fittings design, but it is the endeavor of all Hull Drafting Rooms to standardize them as much as possible. The variations usually being in the size. This is true of Bitts, Chocks, Cleats and many of the small deck fittings which will vary according to the size of rope to be used with them or some other such reason. When it is well designed the fitting will be increased or decreased proportionately in all parts. Because of this standardization, the Pattern Shop is able to store away patterns after use, until they are called for again for similar work on another ship.

Much of the work in shipbuilding is subject to standard rules regarding the structure and plating, especially if the ship is built to "classification" as most of them are; but in the matter of ship fittings the draftsman has a free hand for his ingenuity and inventive inclinations. Fittings are often subject to scientific calculation when proportioning their dimensions and it is the care with which the draftsman works that gives a neat, strong appearance or a rough, clumsy look to his finished product.

Joiner Work

The ship joiner should be more of a carpenter than the regular so-called ship carpenter, as much of his work is fine carpentry or cabinet work.

The ship joinery department has its headquarters on shore with a well-equipped shop for building doors, tables, etc. Much of the built-in furniture is portable and is so designed that it can be shipped in through doors and other openings after the ship has been built. Usually most of this work is in progress while the hull of the ship is being built, and after the launch it is possible to send a large part of the joinery work down to the ship ready for installation. Those parts which must fit within a certain space can be made an inch or so longer and then sawed off to suit the actual distance between bulkheads.

The living quarters of a ship are nearly all furnished with built-in furniture, the chairs and tables being about the only portable articles. Most of the state rooms are fitted with berths under which are a chest of drawers, clothes lockers, and some type of wash-stand. A desk is provided for the officer's rooms. The mess rooms have dining tables and such side tables as may be necessary for serving.

The Wheel House is often completely built in the joinery shop and put on a large truck and taken to the ship after launching. This saves much time, as the wheel house can be so completed that it can be set directly in place.

The inside joinery work, which usually consists of the paneling and other finish, is started in the shop and is put in place as soon as the various other artisans (electrical ventilator, piping, etc.) complete their parts of the fitting and are out of the way. The outside work done by the Ship Joiner is finishing around windows, hand rails, wood ladders and fittings of this type, boat chocks, paneling around front of the Bridge, wood gratings, boxes for deck gear, etc.

CHAPTER XVIII

Engine Room and Engines

The busiest place aboard ship is the Engine Room, for it is here that the energy generated in the Boiler Room is diverted into the main machinery that propels the ship on its voyage or into the various other machines that keep the auxiliary apparatus in operation.

The modern marine engine is now fairly well standardized and of two main types: the reciprocating, in which the motion is back and forth; and the turbine, whose motion is rotary. The reciprocating engine is still, after a great many years, in use and its cylinder, piston, piston rod, slides, connecting rod, crank and crank shaft are quite familiar to most of us. It is the crank shaft of a marine engine that is connected to the shafting in the thrust block. Marine engines of the triple-expansion type with a high, intermediate, and two low-expansion cylinders, are those most commonly employed.

Experiments in recent years have shown that, when its speed was controllable, the turbine was useful for marine propulsion. With the use of the reducing gear, by which the propeller is made to revolve, say once for the turbine's 40 times, it has proved more satisfactory because the arrangement of turbine permits of keeping the weight low, while in the reciprocating engine the weight is considerably higher, above the shaft, due to the weight of the cylinder, also in the amount of space required.

As the propeller must be sufficiently high to be protected from striking any stray logs or other submerged objects, it is customary to have the center raised enough so the blades during the downward swing will not go below the line of keel. This brings the center line of the shafting so high above the tank top plating that high foundations must be made to take the engine bed, rather than drop the base of the bed down, thus increasing the weight to a considerable extent. To keep the Main Condenser in line it is necessary also to have this on a high foundation. Therefore, the grating, called the "walking flat," for the engineers to stand and walk on, is raised some distance above the top of the Inner Bottom. The Auxiliary Machinery is raised to meet the height of the grating, and to suit the best stowage arrangement.

In some of the older shipyards it was the practice to assemble the engine aboard the ship. When this was done, it was customary to strike a center line above the bed plate, indicating the position of the center of the shaft. The columns of the engine were bolted in place, the slides fitted, and the cylinders were set. By plumbing from the top of the cylinder down through the openings, the alignment of the guides was determined. After the crank-shaft was in place, this was rotated through 180 degrees and the opposite positions of the crank gave a dimension square to the shaft. In this manner it was possible to "line-up" the engine, even though the small ships were being built at an inclination on the shipways.

As a result of modern methods of handling heavy weights, it is possible and is largely the practice in modern yards, to build the engine complete in the erecting shop and then dismantle parts of it; carry it down and put it up directly on the ship. When building in the shop the bed plate is placed on a " platen," which is really a platform leveled up in both directions. One advantage in erecting in the shop lies in the fact that it is possible to use the plumbing line in both directions, and the work of the engineers can go on much faster when not surrounded by a lot of men who are still building the ship. The shop, too, has more room and better facilities for doing the work.

Turbines are made in some shipyards, but most of the the shipyards which are installing them in the ships being built in their yards are buying the turbines complete from the manufacturers. The turbines are delivered in advance and stored until ready to be placed in the ship and then fitted on the foundations made ready for them.

The Engine Room of the modern cargo steamship has, beside the main propelling machinery, many small engines and pumps for the work about the ship. These are taken up under the heading of " Auxiliary Machinery."

It is interesting to note the cycle of work done in the Boiler and Engine Room in order to produce the power required to drive the ship. Water is drawn from the Feed-water Tanks to the Boilers by means of the Feed Pump. Here it becomes heated and turns to steam; when it is allowed to escape through the main steam pipe into the Engines. In both the reciprocating and turbine engines the steam enters at one end, called the high-pressure end, and travels through the engine, expanding in successive stages until its energy has been practically all used up. After its work is done in the engine, it is forced out into a pipe and is drawn down by suction into the Main Condenser. Here the steam comes in contact with the cold pipes, becomes condensed, and turns into fresh water again, which on all ocean-going vessels must be carefully conserved.

From the Main Condenser the water is pumped out by the Air Pump to the Feed and Filter Tank. As the engines are well lubricated, some of the oil is absorbed in the steam and carried along with it, so that when the water is run from the Main Condenser, there is more or less oil mixed in with it. As this is detrimental to the inside of the Boilers, it is necessary to remove as much of this oil as possible before the water is again used as feed water for the Boiler. This is the purpose of the Filter Tank. The water enters this tank, which is divided into four or five spaces by diaphragms which are open at the top, allowing the water to flow over. In each of these compartments in the tank there are " Loofa " sponges, or excelsior or some other such material. These are placed to catch the oil and grease, and after they have been in the tank for some time, they are taken out and thoroughly cleansed before being replaced.

Passing from the Hot Well end of this tank (the Hot Well meaning the end where the water is waiting to be drawn off after it has been purified) the water is often passed through another purifier in the pipe line, called the " Grease Extractor " which is a small tank containing two cartridges. These cartridges are really a form to hold Turkish toweling which is wrapped around them and through which the water must pass on its water out of the compartment. These Turkish towels are removed and cleaned in the same way as the sponges of the Filter Tank. After this further cleansing the water is allowed to run back into the Feed-water Tank.

With the ordinary type of reciprocating engine, as the piston can be driven in both directions, it is possible to reverse the action of the propeller by rotating it in the opposite direction with the same engine that is used when going ahead; but from the nature of the turbine engine it is necessary to have a turbine for backing or reversing of the propeller. This backing turbine is on the same shaft as the one for driving ahead and while the ship is going ahead the backing turbine runs free. When it is necessary to reverse the motion of the shaft, the steam is admitted to the backing turbine and the other runs free.

The power developed in the backing turbine is usually about two-thirds of the power of full speed ahead.

"Electric drive" is the latest development for marine propulsion. In this arrangement a high-speed oil-driven engine runs an electric generator. The current is carried by wires to the motor, which is direct-connected to the propeller shaft.

The advantage gained lies in less weight: the engine and generator can be located anywhere and by placing the motor near the After Peak bulkhead, the Shaft Tunnel is not necessary.

CHAPTER XIX

Boiler Room and Boilers

The arrangement of the Boiler Room depends largely upon the type of boilers to be used. For the coal-burning type the coal bunkers are often placed at the side or as a cross-bunker, running from one side of the ship to the other. Where the bunkers run along the sides, the one opposite to the Boiler Room has a door for shoveling coal, and coal in the other bunkers is carried through this door by means of coal buckets which run on a trolley track through the bunkers; the track being covered up and filled in when the bunker is full, but as the bunker is being emptied and becomes free, the coal can be carried to the dumping door, where it is shoveled into the Boiler Room, and is ready for use in the boilers. When this trolley track is not used the coal is shoveled by hand and this is called " Trimming the Coal."

When coal is used, ashes are hauled from below the grate and shoveled into the ash ejector if the ship is at sea. This ash ejector is a large syphon pump carrying water with an open hopper so that the ashes can be shoveled into the hopper, caught by the stream of water and washed up the pipe and out through the ship's side, above the water line. When the ship is in port, the harbor authorities will not allow ashes to be dumped overboard and they usually are carried up the elevator, running up through the ventilator from the Boiler Room, carried out on deck and dumped over the side into a scow.

In the Boiler Room where oil fuel is used, there is a much more neat appearance, the space can be smaller and it is altogether much more satisfactory. The oil is pumped from the Settling Tank by means of the Service Pump. From this pump the piping runs along the top of the boilers with separate branches down to each door of the fire pot, so that each boiler can be independent of all the others.

There are two main types of boilers; the fire-tube and the water-tube. In the first case the fire, which is really hot gases, due to the combustion in the fire box, passes through the fire-tubes and the surrounding water which is in the boiler becomes heated. This type is usually called the Scotch boiler. The diameter is generally more than the distance along the axis. There are three fire boxes, one in the lower part on the center line and the other two just above it on the sides. These fire boxes run to the after end of the boiler, and the gases pass up through a combustion chamber and return to the front end of the boiler again through small tubes. From there the gases pass out into the Uptake and so on up into the Funnel and out into the atmosphere. In this type of boiler, it is necessary to have the top of the fire boxes entirely covered to a sufficient depth with water in order that the crown or top of the box does not become over-heated and collapse. This is the most frequent trouble arising from this type of boiler, other troubles are caused by bad joints or leaky tubes.

The other type of boiler, called the "water-tube boiler," is manufactured in a number of different shapes and combinations. There are at least a half dozen of these, all of which are of good design and durability. In this type of boiler the water is carried in a number of small diameter tubes which are held in position by cast-iron headers. The water is made to flow from one layer of pipes to the next layer above it by means of diaphragms in the layers which separate one compartment from another. By this means of passing the water directly over the fire and through the gases, it can be brought to a high temperature and steam in a shorter space of time than is possible with the Scotch boiler, because of the smaller volume of water. The fire pot, or combustion chamber, in these boilers is as low as possible and the combustion made very thorough by means of baffle plates which hold the gases around the outside of the water tubes until they have lost practically all of their heat, when they pass upwards and out through the Uptake and Funnel.

For some time past it has been customary to use the Scotch Boiler on merchant work and tramp steamers, where there was ample time to get " under way," but for men-of-war where it was often a vital necessity that they should move quickly, it has been customary to use water-tube boilers.

During war time when it is so difficult to get material delivered, etc., in time to complete the ship program, other conditions are arising which necessitate changes from the old order of things, and water-tube boilers are also being used for cargo boats for Government service.

One advantage in using the water-tube boiler lies in the fact that this can be built aboard the ship and finished there complete while the ship is still under process of construction. When using the Scotch boiler, this cannot be done, as that type of boiler must be built in a regular boiler shop, where the necessary rolls and riveting tools can be used.

Access to the Boiler Room is by means of iron ladders from the Upper Deck and by a door from the Engine Room. With the two large ventilators at the forward end of the room giving a good draft of cool air over the heads of the firemen, and with an oil-burning installation of boilers, where the arduous work of stoking the fires is done away with, the work of the firemen in the Boiler Room of a modern steamer is not the struggle for an existence which was so often the case in the olden days.

•

CHAPTER XX

Propellers

Propellers are either made solid (with the hub and blades cast in one solid piece of metal) or they are "built up" (with the hub separate from the blades). Nearly all merchant ship propellers are solid cast iron, and those for the larger men of war are built up with the hub and blades of bronze. In this type the blades have a large screw shape at the base of the blade which threads into a similar shape at the hub. Built-up propellers have the advantage of making it easy to remove individual blades for repair in case of damage, while with the solid propeller, the entire propeller becomes useless until it can be replaced. Nearly every merchant ship carries an extra propeller on deck for use in case of emergency.

As the "Rules of the Road" require vessels meeting head on to pass to the right, leaving each other on the port hand, all single screw ships are fitted with righthand propellers, since these revolve so that the head of the ship would turn to the right with the rudder set amidships, thus helping by a natural tendency what is required by law.

"Twin Screw" ships have two propellers which turn in opposite directions from each other with the tops turning outward from the ship. This working of one propeller against another makes steering the ship easy, and keeps her in a straight course. If both propellers worked in the same direction, it would be necessary to have the rudder set partly across the ship in order to keep her in a straight course, and if it became necessary to steer the ship in that direction the rudder would not have sufficient control of the vessel. It has happened that, as a result of an accident to the rudder, the twin propellers have been used as a means of steering, but it has been found unsatisfactory because it gives the ship an uneven course due to the " wild " steering.

The inside of the propeller hub is bored with a taper and a "key-way." The piece of shafting which fits into the hub is called the " tail shaft." On the after end of this shaft a corresponding taper and key-way is made. At the extreme end of the shaft a smaller, straight part is turned and threaded to take a large nut. This nut is " acorn " shaped with the butt flat to fit against the propeller hub. The forward end of the tail shaft is fitted with a large flange and drilled for bolts.

The shafting ahead of this is called the " tunnel shafting" as it runs through the " shaft alley " or " tunnel," an enclosure built over to protect it from damage by the cargo and to provide access to the shaft at all times, whether the cargo is in the Hold or not. This shafting is in five or more pieces, according to the length of the ship. It is bolted together by heavy flanges and is supported by a " steady bearing " to each length of shaft. These are in halves so the shaft can be set down into them and the " cap " bolted in place afterward. The foundations for these bearings are called " Stools " and they are set up about 4 ft. from the Tank Top. The forward piece of shafting is engaged with the "Thrust Block."

On this piece of shafting there are cast webs, or collars which have been machined to stand at right angles with the shaft, a few inches apart and sufficiently thick to withstand a heavy thrust. The forward end of this shafting is connected direct to the main engine in the Engine Room.

The forward end of the Shaft Alley is widened out to allow room for the "thrust block." There is a bulkhead across the rear of the Engine Room and this is continued around the "thrust recess," as the space for the thrust block is called. A water-tight door is fitted in this bulkhead for passage into the Shaft Alley. As before stated (under the chapter on Bulkheads) the Shaft Alley is built off the center line of the ship to give a passage, generally on the starboard side, for the engineers to inspect and oil the shafting.

The Thrust Block is a heavy, casting having webs corresponding to those on the shaft, as just described, but moved one space so that the collars on the shaft will fit into the spaces on the thrust block. As the propeller revolves and (due to the shape of the blades which cut ahead in the water), gives a thrust forward on the shaft this force is transmitted to the hull of the ship through the thrust block which is firmly secured to its foundation; hence its name. By this means all strain is taken off the engine, whose only work is to revolve the shafting which turns the propeller.

As the front side of the collars on the shaft bear against the back side of the webs on the thrust block it is necessary to keep all bearing surfaces well oiled.

Many times the ingenuity of the engineering staff on some of the ships plowing their way through the different oceans, all over the world, has been taxed to the limit when it has been necessary to make temporary repairs in order to reach land. Sometimes the shafting in the tunnel becomes "tired" and a "fracture" develops. This means "splints" and day and night work until it has been so bound up that it will take the strain of "half speed." Sometimes the propeller nut becomes unscrewed and the propeller drops off and is lost. Then come weary hours and much seamanship on the part of the Captain to hold the ship so the Engineer can lower and ship the spare propeller. Whole books have been written on such accidents and how to overcome them so no more will be said of them here.

CHAPTER XXI

Auxiliary Machinery

Most of the auxiliary machinery is in the Engine Room while two or three pumps are in the Boiler Room. This machinery is necessary to carry on the different functions of pumping water and oil for the different systems described under Piping Systems.

The largest object in the Engine Room beside the main propelling engine is the Main Condenser. The following list of machinery with their descriptions will give a fairly good idea of the work done by them.

Evaporator. All sea-going ships carry fresh water for boiler feed in built-in tanks in the Inner Bottom, but for long voyages the capacity of these tanks is not sufficient to carry all the water needed and an Evaporator is carried in the Engine Room to help out on the water supply. Sea water is run in through this Evaporator, heated to a point of evaporation and then distilled. The distilled water is fed to the Main Condenser.

Evaporator Feed Pump. This pump draws water from the ocean in through a sea valve to the Evaporator. It is located in the Engine Room where it can be easily reached.

Condenser. Every ship carries two condensers, a "Main" and an "Auxiliary." The Main Condenser is used while the ship is under way and the Auxiliary Condenser is used while at anchor in a harbor.

The steam from the boilers is carried direct through main steam pipes to the Engine whether reciprocating or turbine. After the steam has passed through the Engine and done its work, it is discharged out from the Engine through a large pipe to the Condenser.

These condensers are large tanks containing many small tubes through which the exhaust steam passes. During its journey through the tubes it is condensed from the vapor to water, as the outside of these tubes are being kept constantly chilled by sea-water, which is being pumped in from the ocean, forced through the condenser and overboard again.

In the case of the Auxiliary Condenser, the steam from the auxiliary machinery is carried direct to this Condenser rather than the larger one, as the Auxiliary is similar to it and capable of taking care of all the steam used by the auxiliary machinery; but is not large enough to condense steam from the Main Engine.

Main Circulating Pump and Engines. This pump is used to circulate the water through the Condenser, drawing it in through the sea chest in the shell plating of the hull, below the water line, sending it through the Condenser and out through the overboard discharge pipe above the water line. The pump is generally of a centrifugal design, driven by a vertical simplex engine.

Combined Air and Circulating Pump. This pump is used with the Auxiliary Condenser. It has the steam in the middle, the water at one end and the air at the other.

This pump draws from the Condenser into the Feed and Filter tank.

Air Pump. The air pump is made up of a "beam type" with two cylinders. When used with the reciprocating engine, the beam works off the low-pressure cylinder cross head.

Another type of air pump which has been used in the past but is not so common at present, has one cylinder, called the "Simplex" but it is more often used for auxiliary purposes.

Feed and Filter Tank. The ordinary form of feed and filter tanks is a receptacle made into four compartments by means of three diaphragm plates, which are open at the top. Water from the Air Pump is pumped into the end compartment, across the tops of the diaphragm plates into the other end compartment and is then carried off through the lower part of this end of the tank.

In these compartments there are placed "Loofa" sponges, excelsior, or some such material, which catch the grease off the water.

The purpose of this filter tank is to purify the water as much as possible from the grease which has collected during the passage of the steam through the engine, so that when the water has returned as new feed, it will not be greasy water which is run into the boiler.

Feed Water Heater. This heater is used to raise the temperature of the feed water during its passage from the water tanks to the boiler. It consists of a tank containing coils of pipe through which steam is passing. The feed water passes through the tank around the outside of these coils and is heated by coming in contact with their hot surface.

Fresh Water Pump. This is used for pumping fresh water for drinking and culinary purposes to the different parts of the ship direct or to a Gravity Tank placed on the Boat Deck for this purpose.

Fire and Bilge Pump. This pump is used in case of fire for pumping direct from the ocean through the system of fire hose which carry the water to all parts of the ship. The bilge water is aired through this pump by reverse action and it is possible to pump out any part of the ship's bilges through this pump, the discharge going aboard.

Ballast Pump. This pump is used for moving water ballast from one Inner Bottom compartment to another or for emptying any or all of the compartments, as may be necessary.

Sanitary Pump. This pump is used to force salt water through the sanitary system of the ship, supplying all toilets, deck washing gear, etc.

Lubricating Oil Pump. This pump is used to keep up the constant supply of oil from the lubricating oil tanks to the different bearings of the engine, which require constant lubrication while running.

Oil Cooler Pump. The oil cooler is a receptacle used for passage of lubricating oil while it is being cooled by means of circulating cold water through a portion of the cooler.

Ice Machine Circulating Pump. This pump is used to maintain circulation of air through the condenser of the ice machinery.

Main Feed Pump. This pump is used to draw water from the feed tank to the boiler when the ship is at sea and the main engine running.

Auxiliary Feed Pump. This is similar to the Main Feed Pump and is used, when the ship is in harbor, for the auxiliary machinery. It is run by steam, the same as all the other pumps and is of a duplex type.

Fuel Oil Transfer Pump. This pump is used to force oil from one compartment into another; as the different compartments are emptied and others are to be filled up.

Fuel Oil Service Pump. This is used to pump directly from the Settling Tanks to the Fuel Oil Heater.

Fuel Oil Heater. During its passage through this heater, the oil is raised in temperature until it becomes vaporized and is forced out in the form of gas through the spray nozzles in the boiler.

CHAPTER XXII

Piping Systems

On shipboard it is necessary to convey water, oil and steam for various uses, and for this purpose pipes are employed. Various systems of piping are common to all ships and they differ only in intricacy according to the requirements of the various types of ships. Men-of-war have a most complete and complicated system of piping, while the system aboard transports and other passenger carrying ships is rather less complicated. The ordinary cargo ship (in which we are interested) requires practically a minimum of piping outfit.

The piping lines used for different purposes vary both in kind of material, size, styles of joints, and in other ways. The piping of a ship forms an important item in its construction work, and is carried on in conjunction with the building of the ship structure.

All the systems have forced circulation driven by pumps in the Engine Room controlled by the engineers. Some systems are so connected that if the pump for that system breaks down it can be connected to another pump and function just as well.

In the Boiler Room, against the forward bulkhead there is generally a row of manifolds which control the piping for the fuel oil tanks. In the Engine Room, against the forward bulkhead, out near the ship's side, there are two rows of manifolds and a third lot in the thrust recess, which control the other piping.

A manifold is a cast-iron fitting with numerous valves in a row. The piping is connected to this manifold and by opening a certain valve and closing all others except one, the oil, water, steam or whatever is being passed, can be sent in an entirely new direction. By opening and closing certain other valves, many different combinations can be made. For instance, if one pump is to operate for a number of pipe lines, they will all lead into a manifold and then any one of them can be connected to the pump by means of the various valves.

Bilge Piping is one of the important systems in all ships. "Bilge" water means any water in the bilges of the ship. It may be clean but more often is dirty. Any water in the Cargo Holds, through rainy weather when they are open or any liquids due to breakage or otherwise which may run out and onto the Tank Top, mingling with the moisture there or any water in the hold due to a leak, in fact anything of a liquid nature is called "Bilge Water."

At the end of each Inner Bottom compartment there is one floor space which is separate from the tank and into which the Bilge Water can drain. These spaces are called "Sumps," "Sump Tanks" or "Wells." They usually run from the Center Line Keelson out to the side of the ship, from the Tank Top to the Bottom Plating, for one floor space.

The bilge pipe starts in one end of these sumps, there being one pipe for each Sump. From the Sumps the pipes run along the top of the Inner Bottom close to the ship's side, bending out in a loop to clear the brackets of the web frames. The piping runs through the watertight bulkheads to the manifold in the Engine Room from where it is carried to the Bilge Pump and then overboard or to the Fire Main.

The Bilge Piping in the after end of the ship runs from the Sump Tank up on to the Tank Top and along the Shaft Alley, under the Shaft Stools, on the center line of the ship. The manifold for this piping is in the Thrust Recess at the after end of the Engine Room.

The pipe to the Sump Tanks has a check valve in it just before reaching the tank. The valve can be operated at the valve or on the weather deck, by means of a rod. This is necessary as many times the valve must be operated when the Hold is full of cargo and the valve can not be reached.

The Fire and Bilge Pump is in the Engine Room and is connected to both manifolds at the sides, near forward bulkhead of Engine Room and to the one in the Thrust Recess. This pump will draw from the sea and from the Bilge Manifold, and will discharge to the Fire Main and overboard. (This combination of having one pump for the bilge water and for water for extinguishing fires is common on shipboard.)

Ballast Piping is run through the compartments in the Inner Bottom and is used to convey oil or water to different compartments or to the "Settling Tanks" before the oil goes to the Boiler Room. Ballast Piping must be installed when the ship is being built as the piping runs through lightening holes in the solid floors and through brackets floors. At the oil- or water-tight floors it is flanged and holes are cut just the size to allow it to pass, the flanges making a tight joint at the floor. As the piping is in long lengths, they must be slipped in through the floor holes when floors are being laid, but the pipes are not connected up until later when the Piping Department is ready to do the work. Ballast Piping obtains its name from the work it does; to carry water ballast or oil, that may be used partly for ballast, from one compartment to another.

From each compartment there is a pipe leading to the manifold near the forward bulkhead of the Boiler Room and from there it is carried to the Ballast Pump in the Engine Room.

Both the Fore and After Peaks are pumped empty through the Ballast Piping. Pipes are carried to the manifold the same as for any of the compartments.

The Ballast Pump will draw from the sea, from the Ballast Main and from the Bilge Main and will discharge to the Settling Tanks, overboard or to the Fire Main. (This combination of drawing also from the Bilge and Fire Mains is a common arrangement as this pump can then be used in case of any trouble with the other.)

Fire System. This piping is independent of all others as it is designed to be used in an emergency and must always be ready for such use; hence it must not be used in connection with any other work so that, by chance, it might be in use at a time when it was needed to extinguish a fire.

This piping runs direct from the pump up through the Engine Hatch and branches out on the weather decks with leads running both forward and aft. There are outlets placed at advantageous points where canvas hose can be attached and it is generally so arranged that these are near enough so that any two hose can play on the same part of the ship. In other words, so that if there is trouble with one hose the other one can do the work.

The outlets are brass couplings fitted with a valve and the canvas hose is stowed within easy reach of the coupling so it can be ready for immediate use.

One lead of iron piping runs direct from the pump to the Fore Peak Tank and another lead runs to the After Peak Tank. These tanks are filled by this system (and emptied by the Ballast Piping).

As explained under Bilge Piping the pump which does the work for this system also acts for the Fire System as these two systems would not be working at the same time or if the Bilge System were in use it could be quickly turned over into the Fire Main.

This pump is always known as the Fire and Bilge Pump and will pump the ship free of Bilge Water, or water due to a leak in the hull or pump water in from the sea and throw it onto a fire.

There is another system of piping which is not, strictly speaking, part of the "Fire Main," but is used for the same purpose. This is a series of pipes carried from the Engine Room to different parts of the Cargo Holds with open ends on the pipes. Whenever there is a fire in the cargo the hatch covers can be left on the hatches and steam sent into the hold in which the fire is burning. The steam fills the air and smothers the fire. These pipes are called "Smothering Pipes" for this reason.

Sanitary Piping. This is used for carrying sea water for "housekeeping" purposes. The water direct from the sea is used in flushing water closets, washing down Galley floor, decks, and any purpose where water would be used that need not be fresh water. (Fresh water is carefully kept and used on board ship for cooking, drinking and bathing.) The piping is also carried to the "Oil Coolers" in the Boiler Room where it acts as a cooling agent in the water jacket around the oil.

The Sanitary Pump will draw from the sea and discharge to the Sanitary System, Water Service and Oil Coolers.

Main Steam Piping is the large diameter, heavy piping from the boilers to the main propelling engine, whether reciprocating or turbine. The branches from the boilers are run to one common pipe in the Boiler Room and through the bulkhead to the engine. Valves in the branches make each boiler independent of the others. The main steam valve shutting off the flow of steam through the pipe to the engine. Leads of these pipes vary with the arrangement of the Boiler and Engine Rooms so no attempt will be made to describe the layout of piping.

Auxiliary Steam Piping is the multitude of small pipes which serve the auxiliary machinery. Each system of piping takes its name from the pump which it serves. To an experienced workman installation of the piping in the Engine Room of a ship is an arduous job and to one who rks on it for the first time it is very baffling and diffi- t to understand. After the piping is out of the Engine om and on its way to the end of the line the work is re simple.

The arrangement of this piping is laid out in the Draft- ; Room of the Engine Department and is closely fol- ved by the workmen on the job. In the Navy the ing is painted certain standardized combinations of or on the joints so that men who are working the ship d were not on board when it was installed will be able trace the pipes through the ship when making any airs or alterations.

The principal piping for the auxiliaries are the follow- :

Fire and Bilge Pump
Ballast Pump
Sanitary Pump
Fresh Water Pump
Evaporator Feed Pump
Ice Machine Condenser Circulating Pump
Combined Air and Circulating Pump
Main Circulating Pump (and engine)
Main Feed Pump
Air Pump
Auxiliary Feed Pump
Lubricating Oil Pump
Oil Cooler Pump
Fuel Oil Transfer Pump (In Boiler Room)
Fuel Oil Service Pump (In Boiler Room)
Main Surface Condenser

Auxiliary Surface Condenser
Evaporator
Feed and Filter Tank
Feed Water Heater.

As all of these systems of piping are in the Engine Room, except those noted as being in the Boiler Room, and some Engine Rooms are about twenty-five feet long and fifty-five wide, there would seem to be considerable piping in a small space.

Among the auxiliary machinery there is an important system not yet mentioned; it is the " Steam and Exhaust " for the " deck machinery." This includes the Steam Windlass on the Forecastle Deck, the Capstan on the Poop Deck, the Steering Engine under the Poop Deck and all the Cargo Winches at the hatches, for handling the Cargo Booms. All of this piping is well insulated to protect it in cold weather. Steam piping for the signal whistle and siren at the funnel is a separate lead of small sized pipe.

Fresh Water Piping. This system draws from the Fresh Water Tanks and from the Reserve Feed Tanks.

The Fresh Water Tanks are generally located in one corner of the Engine Room. They are two, large, square tanks of steel plate, well riveted and caulked tight. They have swash plates to prevent undue motion of the water when the ship is rolling in a seaway and which also help tie the sides of the tank together, to prevent bulging of the flat surfaces. These tanks are placed on strong foundations and are well secured, to hold them in place.

The Reserve Feed Tanks are those compartments in

the Inner Bottom, usually directly under the Engine Room, which are used for storing fresh water. This water is used for feed water to the boilers as needed. It is generally as clean as the water in the tanks in the Engine Room. To prevent rust from the plating getting into the water the inside of these tanks is coated with neat Portland Cement. In order that the daily service of fresh water need not depend on a pump pressure, a Gravity Tank is installed on the highest deck, water is pumped directly there from the water tank in the engine room and the water is supplied from the Gravity Tank to the different faucets in Living Quarters and Cook's Galley. For drinking water for the crew a "Scuttle-butt" is generally placed on the Upper Deck and is connected to the Fresh Water System. It is a small tank set up on a stand, with cups for drinking.

All fresh water discharge, that water which has been used, drains overboard through a scupper in the ship's side.

Steam Heating System. All Living Quarters, as mess rooms, state rooms, wheel house, washrooms, etc., are heated by steam, as the ships must be comfortable during the extreme winter weather on the North Atlantic Ocean.

The radiators are made up of "coils," usually 1 in. brass pipes with return bends so they are flat and can be hung on the bulkheads, to save space.

The Cook's Galley has a steam-heated table for keeping food warm, steam coil in the bottom of the sink for heating dish water and urns for heating water for coffee, etc. The baths and showers have a steam heater; a large pipe which contains steam and through which the water pipe is passed, the water becoming heated while in transit.

All steam supply and return pipes are well insulated. The drain pipes from the heaters are all trapped and return to the drain tank in the engine room.

One of the important leads of piping is the steam coils in the Inner Bottom. The cold water which a vessel passes through during voyages in the North Atlantic in the Winter chills the hull to such an extent, below the water line, that the fuel oil in the Inner Bottom will become thick and refuse to flow readily through the feed pipes to the Settling Tank, preparatory for use in the boilers of an oil burning steamer. It is necessary to have means for keeping this oil sufficiently warm to flow. Radiator coils are fitted in the end of each tank, near the suction pipe for the oil, and are supplied with steam the same as any radiator in the Living Quarters. They are drained by an extra pipe. By means of these coils the temperature of the oil can be governed so that it is available at all times.

CHAPTER XXIII

Hull Engineering

Steering Gear

Steering of the ship is usually done by a small hand wheel located in the Wheel House on the Bridge. Often a second wheel is fitted on top of the Wheel House, vertically over the other, and is connected to it by a rod which can be disconnected within a few minutes if so desired.

Directly attached to the head of the rudder post there is a yoke with two arms, standing at right angles with the keel. At the outer end of these arms connecting rods run forward to a sleeve which travels fore and aft according to the rotation of the threaded shaft on which the sleeve rides. The arm on the yoke, connecting rod and sleeve are in two sets which are built side by side.

The forward end of the threaded shaft is turned by means of a gear wheel. This gear wheel interlocks and can be turned by a combination of gears which lead from a small, steam "steering engine," or can be operated by a double hand wheel which can be handled by four men.

The steering engine can be operated by either of the steering wheels, on or in the Wheel House and by a small wheel close by the engine. All of the steering gear machinery is located on the deck just forward of the rudder stock, and is enclosed by a light, steel bulkhead.

The small steering wheel and the double hand wheel, aft, are for an emergency in case the forward wheels break down. Should the steering engine itself be in trouble the ship could still be managed as a Spare Tiller is fitted on top of the Rudder Stock, above the yoke. In the end there are two eyes for attaching two sets of rope tackle. By removing the pins which hold the connecting rods to the arms of the yoke and swinging the rods out of the way, the steering tackle being hooked into pad eyes on the ship's side (put there for the purpose), it is possible to steer the ship until she reaches port or until the necessary repairs have been made on the steering engine.

Communication between the forward steering wheels and the steering engine is maintained by two pipes which lead from the base (or standard) of the wheel in the Wheel House, down, under the decks and aft to the steering engine. These two pipes are filled with a solution, often glycerine and water, under pressure, and by turning the forward wheel, plungers in the standard are moved and, by increased pressure in one pipe and decreased pressure in the other, corresponding plungers in the steering engine are affected and when these move they operate the slide valve in the steam chest and the steering engine acts to turn the combination of gears and thus swing the rudder to port or starboard.

Ventilation

Ventilation in battleships and large passenger steamers is quite a problem, but for the type of cargo steamer which is being described the ventilation is a simple matter.

All the living quarters are above the main weather deck except the Engine and Boiler Rooms, and receive all necessary ventilation through doors and air ports.

The Engine Room often has four cowl ventilators fitted, one in each corner of the Engine Hatch, standing above the skylight enough to have an unobstructed draft.

The Boiler Room is fitted with two cowl ventilators, of larger diameter than those in the Engine Room. They are fitted in the two forward corners of the boiler hatch and stand above the Fidley Hatch high enough to have a good draft. The cowl hood of all these ventilators is built separate from the ventilation trunk pipe (which is round in section) and rests on top of the trunk as a guide, so the cowl can revolve. On the outside of the cowl, below the throat and just above the trunk a gear rack is fitted to engage with a pinion gear wheel. The handle to operate the wheel is below in the Engine or Boiler Room, directly under each ventilator trunk so the men in the rooms below are able to turn the cowl up on deck until it catches the direction of the wind and gives a good down draft to the rooms below.

The ventilators in the Boiler Room are made larger in diameter as they are also used as elevator shafts. The Boiler Room has ash buckets which are used to raise and dump ashes over the side into the dump scows while in harbor, if the steamer is a coal burner. Oil-burning ships are also often provided with these buckets.

This bucket is hoisted by a rope over a drum in the ventilator trunk which is turned by a crank on the outside of the trunk. Light angle bars are used as guides to keep the bucket in center of the trunk. A door is in the side of the trunk just above the weather deck, with top of the door a little below the rope drum. The turning gear for the cowls is independent of the elevator and the cowls can be operated in the same way as those over the Engine Room.

The ventilation trunks are built down until they stop a little above a man's head. In this way all the down draft is carried to the Engine and Boiler Room floors where it is needed.

The top of the Engine Room Casing ends in a skylight, built with a number of small shutters, hinged at one end, with a bar to hold the other end open at varying heights. These shutters have glass ports to admit light. The shutters can be opened and closed by a gear operated from below. When an exhaust is wanted to draw out the hot air, the skylight shutters are opened according to the wind; with fresh air pouring down through the ventilator trunk from the cowl, a fine draft is created. The top of the hatch on the Boiler Room is called the " Fidley " and it is cut out for the funnel and the two ventilator trunks to pass through.

In addition, some Fidleys have rectangular openings as large as possible, the size being limited by the framing of the hatch. To prevent accidents, these openings are protected by gratings of round iron bars and having no fixed covering, they can be covered with tarpaulin (canvas) during stormy weather, if so desired. These openings serve the same purpose as the skylight shutters over the Engine Room, being an opening for the exhaust for hot air.

Electric Lighting

The ordinary layout for the electrical installation places two marine type generating sets, direct connected to a small steam turbine on the "Dynamo Flat." This is usually a small platform built half way up, on one side of the Engine Room, where it is easily accessible and out of the way.

Practically all that is required of this outfit is to illuminate the ship as all power is steam driven. The slate switch board is located in the Engine Room where it can be attended by the Engineers.

The "Side Lights" (port and starboard), Stern and Anchor Lights are built to use either electric or oil lamps but the Range and Masthead Lights are electric only.

All Living Quarters as Mess Rooms, State Rooms, Galley, etc., Engine and Boiler Rooms, Shaft Alley, Paint Room and Store Rooms are lighted by incandescent lamps. The weather decks are lighted by means of single lamps attached to the outside of the deck houses.

Work about the ship and in the Cargo Holds is done with "Cluster Lights," there being a number of lamps attached to a single holder and protected by a large reflector which throws all the light in the direction desired. These "clusters" are on long, portable electric cables so they can be placed where work is going on. The generators are designed to stand a sudden change from no load to full load and vice versa, so it is not necessary to ever think of them when using the electric current.

CHAPTER XXIV

Engine Dock Trial

After the boilers and main steam line have been tested out by applying the required hydrostatic pressure, the water is blown out; if the boilers are of the Scotch type, the local inspectors enter the boiler to see if any distortion has taken place while the pressure was applied.

If the boiler is of the water tube type, the water is blown down below the steam drum, and the Inspector makes an inspection of the steam drum only. The boiler is then closed up and the water pumped up till it shows about 3 ins. in the glass. The fire bars are then primed with a layer of lumpy coal, about 3 to 5 ins. thick, and the necessary wood thrown on top to ignite the coal.

If the Donkey Boiler, or any other boiler on the vessel, is under steam, some hot coal can be taken from the furnace, instead of using wood, to ignite the green coal.

The air cock, which is provided on top of the boiler, and the gauge cocks, which are at about the water level, should be opened, so that as the water expands and steam is formed, any air in the boiler will be forced out.

Scotch boilers are usually equipped with Hydrokemeters, or if the piping is so arranged the water can be circulated by taking it out of the bottom blow connection, and returning it to the boiler through the feed line. As soon as fires are started the auxiliary feed pump, which is usually connected for the circulation, should be started. This can be done by connecting the air hose and breaking the exhaust line, and the pump can run until there are 30 or 40 lbs. of steam on the boiler.

When steam is formed so that it can be observed escaping from the air cock, the small gauge cocks which were left open should then be closed and the steam allowed to rise gradually to the required pressure for "cutting in."

If the boiler is of the "water tube" type, it is not necessary to circulate the water, and Hydrokemeters, or other methods of circulation, are never used.

On boilers of the Scotch type, provided with some method for circulating water, steam can safely be raised in from three to five hours. On the boilers of the water tube type this can be done in much less time—say from an hour to an hour and a half, if there is a proper method for forcing the fire.

In new vessels, where steam is being raised for the first time, it is best to "cut in" the boiler; that is: open up the top valves to the auxiliary machinery when the steam has reached sufficient pressure to work the auxiliary machinery, which would be around 75 to 100 lbs.

In cutting in the steam line, the valve should be only "cracked" off the seat, and the steam allowed to rise in the steam lines gradually, as a sudden opening of the valves would tend to lift the water from the boiler, when it would pass over the steam lines into the auxiliary machinery.

When auxiliary machinery is run for the first time (and this is also true of dock trials) the condensed water should not be allowed to return to the boilers, as in manufacturing (in assembling the piping) a great deal of oil and grease is used. By letting this water run into the bilge, the accumulated oil would not return to the boilers and cause injury. This is generally done by breaking the discharge pipe from the main and auxiliary air pumps, thus letting the water flow to the bilge and " make-up " feed for the boilers is taken from the shore lines.

After all of the different pumps (dynamos, steering engine, winches, and other auxiliary machinery) have been tried out under steam, and it is found that they function satisfactorily, the vessel is then ready for dock trial.

Before the propeller is turned over, it should be seen that sufficient lines are out fore and aft, and also " breast " lines, so that the engines can be worked in both the ahead and stern motion, without moving the ship more than a small distance from her position at the dock. When the word is given to get under way, the first precaution to be taken is to give the engines a complete revolution with the jacking engine, and the engineer noting that there is none of the oil piping, or other parts that foul anything. Then the jacking engine gears should be taken out and secured; the main sea valve injection should be opened, the overboard discharge valve opened and the main circulating pump started up.

Air cocks are usually provided under the condenser heads to let out the air, so that water will pass through all of the tubes, which would be pocketed if there were not an air vamp.

After it is found that the main circulating pump is functioning properly and the main and auxiliary feed pumps will deliver water, the main stop valve should be cracked on the engine, allowing steam to pass up to the throttle valve. If there is a drain on the main steam line this should be opened so that whatever condensed water is standing in the pipes may drain off and make sure that nothing but dry steam reaches the engine.

The reversing engine should be worked a few times in both directions, if the engine is of the reciprocating type, and then crack the by-pass valve to the high, intermediate and low cylinders, if it is of the triple expansion type, leaving all drains open.

The reversing engine should be worked at intervals (every minute or two) so that the steam passing into the valve chest and on into the cylinders, will have ample chance to reach both sides of the piston, and have all parts of the cylinders warmed up. After the engines have been warmed sufficiently, the throttle valve should be cracked and the engines allowed to make a half turn, and then brought into the reverse motion with reversing engine, or (what is commonly termed by engineers) the engine should be " rocked." After this has been done for several minutes, the engine should be then allowed to make a few revolutions in the ahead motion and all bearings and other parts watched so that they should not be warming up or becoming overheated. The number of revolutions of the engine should gradually be increased until the engine is developing the proper horsepower, and should be run at this speed for from two to three hours. If any of the bearings become overheated, or any adjustments are necessary, the engine should be stopped and the proper adjustments made.

In order to run a satisfactory Dock Trial, it sometimes is necessary to run for a short time and stop the engines and make the necessary adjustments for whatever trouble may develop.

When the engines have been run sufficiently to determine their condition and all is found to be satisfactory, the steam is allowed to fall and the engineers turn their attention to "slicking up" their quarters. From now on until the time arrives when the ship is fully equipped and ready for sea the Engine Room Force is becoming accustomed to the ship and all layout of piping, etc.

(Thanks are due to HADLEY F. BROWN for assistance in preparation of this chapter.)

The progress of this ship has been traced from the time the first keel plate was laid, through the construction of the hull, the launching and fitting out of the propelling machinery and other equipment.

The description has been brief in many parts, of necessity, due to space allowed in this little book. For more detailed description of the engines and boilers the reader is referred to the technical books which are written exclusively on those subjects. The problems of design which are met in the Drafting Rooms of the shipbuilding company that is building this boat, have not been referred to at all, as that is not the purpose of this volume. For such information one should obtain a book giving the work in "Theoretical Naval Architecture," as this description has been written with a view of showing the average shipyard worker some of the things which are going on around him and with which he may be unfamiliar. If it has been of any such help the writer will feel well rewarded for the evenings spent in preparing these pages.

As a last word, don't forget SAFETY FIRST, for yourself and for your fellow workers, for your family and for theirs.

Never look aloft while walking on deck or staging. If it is necessary to look upward, stand still while doing so.

Never place wrenches or other tools where they may fall—think of the men below you.

"Eternal Vigilance is the Price of Safety."

SHIP NOMENCLATURE

NOTE.—As this volume deals with "Practical Shipbuilding" of modern vessels, the large majority of which are steel steamers, it is the purpose of this Glossary to include only those ship terms or names of structural parts which are commonly used in the shipyard during the construction of the vessels.

Hence, structural parts of wood ships, rigging of "fore and aft" and "square rigged" vessels; scientific terms used by the naval architect when designing the hull and machinery, have been omitted.

A

Access Hole. An entrance to a compartment.

Accommodation Ladder. A ladder or stairway which can be suspended from ship's side to permit a person or persons to board a ship from a small boat.

Aft. Position between amidships and stern of ship.

Aft End. Position at or near stern of ship.

After. Nearer the stern.

After Peak. Aftermost tank or compartment of a ship, used for trimming the ship.

After Peak Bulkhead. Last water-tight bulkhead of a ship.

After Perpendicular. A vertical straight line at the after edge of the rudder post.

Air Duct. A passage for carrying air; often running from a lower deck to a weather deck.

Air Hammer. Hammer driven by compressed air for riveting. Sometimes called an air gun or "gun."

Air Port. A circular opening in the ship's side.

Alignment. The lining up, exactly straight, of a keel or propeller.

American Bureau of Shipping. An Amercian organization for classifying ships. Similar to "Lloyd's Bureau of Shipping."

Amidship. Section of a ship half way between bow and stern.

Anchor. A heavy steel device that is attached to the end of a large chain or hawser to be dropped to the bottom for holding the ship in position when not alongside a dock.

Anchor, Bower. An anchor stowed in the bow of the ship.

Anchor, Kedge. An anchor used for "kedging," or hauling a ship along a channel.

Anchor, Selfstowing. An anchor without stock, whose flukes are hinged and fold up against the side of the ship as the anchor is hauled up into the hawse pipe.

Anchor, Stream. A medium-sized anchor, about twice the size of a kedge anchor.

Angle Collar. Flange made of angle iron.

Angle of Rudder. The angle a rudder makes with the fore and aft center line of the ship.

Anneal. To heat and cool metal for the purpose of softening.

Anti-Fouling Composition. A patent paint applied to the bottom and sides of a ship to prevent marine growth below the waterline.

Apron Plate. A bulwark plate at stem head, extending aft a short distance for housing " bow chocks " also to assist in keeping tha forecastle dry.

Ash Ejector. An inclined chute running from the stokehold, up and out through the ship's side above the waterline, to eject ashes.

Assemble. To fit two or more parts together, forming a section of a larger part.

Athwartships. Across the ship, at right angles to the keel.

Auxiliaries. Various winches, pumps, motors, and other small engines required on a ship.

Auxiliary Foundations. Supports for pumps, distillers, condensers, etc.

B

Backing Angle. An angle fitted on the back side of another angle to reinforce same.

Ballast. Substance carried by a ship to insure stability.

Ballast Tank. Water-tight compartment to carry water ballast.

Barge. A cargo vessel with no means of propulsion.

Batten. A thin strip of wood used in measuring or making templates. (See also Cargo Batten.)

Beam. An athwartship member supporting a portion of a deck. Also the width of the ship.

Beam Bending Machine. A powerful bending machine made with a yoke and plunger for straightening or curving deck beams.

Beam Bracket. Plate at ends of beam, riveted to beam and frame.

Beam, Cargo Hatch. Portable I beams supported by the hatch coaming.

Beam, Half. These are in way of cargo hatches and extend from the hatch to the ship's side.

Beam Knee. Bracket made by splitting and bending the end of channel to form a knee brace.

Beam Mold. A wood pattern made in the Mold Loft, giving the shape to which the beam should be bent.

Beam, Strong. These are at forward and after ends of cargo hatches and take the weight of the fore and aft hatch girders.

Beam, Whole. This are continuous from one side of ship to the other side.

Below. Below a deck or decks (corresponding to "downstairs").

Bending Rolls. Large machine used to give curvature to plates.

Bending Slab. Heavy cast-iron blocks arranged to form a large solid floor on which angles are bent.

Berth. A place for a ship; sleeping accommodations.

Bevel. The angle between the flanges of a frame and other member. (When greater than a right angle, "open bevel"; when less, "closed".) To change the angle of an angle bar tomake it fit in a certain place.

Bevel Board. A board used for marking off with numbers to give the bevel from one frame to another.

Bevel-faced Holding-on Hammer. A heavy hammer with sloping face for holding against a rivet.

Beveling Machine. A portable machine on wheels which is carried in front of an angle furnace and through it the angle is run as it is drawn hot from the furnace. The beveled slot through which the angle travels gives the required shape to the flange.

Bilge. The rounded portion of the hull between the side and bottom.

Bilge Crib. An open cribbing of blocks placed under bottom of ship to support the bilges during construction, or while in dry dock.

Bilge Keel. A fore-and-aft plate fitted to the outside of the shell plating running along the bilge, used to prevent excessive rolling of the ship.

Bilge Pump. Pump for removing waste water from bilge of ship.

Bilge Strake. See "Strake."

Bilge Stringer. A stringer inside and near the bilge of the ship.

Bilge Water. Water which accumulates in the bilge of a ship.

Bitts or Bollards. Heavy steel castings fitted to the weather deck for securing mooring lines or hawsers.

Bitumastic. An elastic cement covering for inside of compartment, decks, etc., to prevent material from rusting and prolonging life of same.

Boat Chock. A wooden or metal piece cut to fit the under portion of a life boat while resting on the deck of a vessel.

Body Plan. A view showing curves of the frame lines at any point in the ship. Frames forward of amidships are on the right of the center line, aft of amidships, on the left of center line.

Boiler Saddles. Supports for boilers (foundations).

Bollard. Ventilating. A bollard or bitt with hollow center used for passage of air to compartments below.

Booby Hatch. See "Hatch."

Boom. A long round heavy spar, pivoted at one end, generally used for hoisting cargoes, etc.

Boom Crutch. Support for booms when same are not in use.

Boot-topping. Sometimes this is formed by the carrying up above the water line of the same color paint as used for the bottom plating. On other boats it may be formed by a band of color separate from that used for the bottom and that for the "top sides" (above the water line). This line of paint is only a short distance above the water and it is designed to give a jaunty effect to the appearance of the ship.

Bosom Piece. A short angle connecting the ends of two angles.

Boss. The curved portion near the stern post, of the ship's hull around the propeller shaft.

Boss Plate. Plate bent to fit around the boss in stern casting at the propeller shaft.

Bottom. Portion of the hull below the bilge.

Bounding Bar. Angle iron fitted around the edge of a plate or plates of bulkhead or deck to make a water-tight connection.

Bow. The forward end of a ship.

Bow Port. A port in the bow of a vessel to allow the stowage of very long pieces of timber, etc. Generally used in wooden schooners.

Bracket. A small plate used to connect two or more parts, such as deck beam to frame, frame to margin plate, etc.

Breaching. A Y-shaped sheet steel casing which connects the boilers to the funnel. It forms part of the "uptake."

Breakwater. A structure built on the forward end of a flush deck ship to protect the crew from waves which might wash over the bow.

Breast Hook. A plate structure fitted inside the hull near the bow to give local strength to the shell plating.

Bridge. The athwartship platform above the weather deck from which the ship is steered, navigated, etc.

Bridge Deck. See "Deck."

Bridge, Long. Built-up span connecting forward deck houses with after deck house.

Broken Stowage. Projecting parts in a cargo hold which interfere with the solid packing of cargo.

Brow (Ramp). This is an inclined walk formed of plank and uprights and fitted with cross pieces of wood to prevent slipping. Generally used for a passage for the workmen from the ground to the Upper Deck, for access to the ship while building.

Bucker-Up. The man who "Holds-on" against head of rivet.

Buckled Plate. A plate which is bent out of shape.

Building Slip. Place where the ship is built, before launching.

Bulb Angle. An angle with one edge rounded.

Bulb Plate. Plate reinforced at one edge.

Bulb Toe. Bar reinforced at toe of standing flange.

Bulkhead. A vertical partition corresponding to the wall of a building, extending either athwartships or fore-and-aft, and separating one compartment from the other.

Bulkhead Boundary Bar. The outside bar of a bulkhead where it connects on to the decks and shell of the ship.

Bulkhead, Cabin. Wooden partitions to form cabin compartments.

Bulkhead Casing, Boiler. Bulkheads which run from one of the lower decks to the weather deck, forming a closed-in space or large trunk.

Bulkhead Casing, Engine. Similar to the Boiler Casing except that it extends around the Engine Room space and is finished by a skylight on the top.

Bulkhead, Collision. The foremost bulkhead, which would serve as the front of the ship if the bow was destroyed by collision.

Bulkhead, Corrugated. A system of using corrugated metal for light partition bulkheads and omitting the stiffener angles.

Bulkhead Doors. Either vertical—watertight—sliding doors; hinged watertight or hinged non-water tight doors in a bulkhead.

Bulkhead Liner. A diamond-shaped plate against the connecting edge of a bulkhead to an outside shell plate.

Bulkhead, Partial. A bulkhead which does not completely extend to the deck above.

Bulkhead, Screen. A light-weight bulkhead between Engine and Boiler Rooms.

Bulkheads, Side Bunker. Longitudinal bulkheads placed in way of machinery spaces.

Bulkhead Sluice. A small opening cut down to top of the bulkhead angle connecting to tank top; fitted with valve, and worked with a rod from any deck above.

Bulkhead, Stepped. A bulkhead in which the upper portion does not come vertically in line over the lower section.

Bulkhead Stiffeners. Angle or channel bars fastened to the plates of a bulkhead to stiffen it.

Bulkhead, Wash. A partial bulkhead in an oil or water tank to prevent the splashing of the oil or water when the tank is not full. (Same as Swash Plate.)

Bull Dozer. Same as Beam Bending Machine.

Bull Riveting. Riveting with a compressed air or hydraulic plunger riveter.

Bulwark. Shell plating extending above the top deck of a ship.

Bulwark Stays. An inclined brace from deck to inboard side of bulwark plating to stiffen the bulwarks.

Bumper. See "Fender."

Bunker. A compartment used for the stowage of coal r other fuel.

Bunker Door. A small, vertical, sliding door in a bunker ulkhead to admit coal to stokehold.

Bunker, Pocket. A conduit for passing coal from the pper bunkers to the stokehold.

Bunker, Reserve. A small watertight athwartship or cross" bunker used either for coal or cargo.

Bunker Stays. Angle struts connecting frame of ship to unker bulkheads to stiffen the bulkheads.

Buoyancy. Ability to float, or the difference between he weight of the ship and the upward force of the water hat it may displace.

Buoyancy, Reserve. The volume of the water-tight hull bove the load water plane.

Burr. The metal bent outward from around the far dge of a punched hole.

Bushing. A metal or lignum vitæ collar used around a evolving shaft to take up the wear.

Butt Joint. See "Riveted Joint."

Butt Strap. A small plate or other member used to onnect the two parts of a butt by overlapping each.

Butterfly Nut. A nut with two large "wings" for tightening by hand.

Buttock. The convexity of a ship aft, under the stern.

C

Cable. Manila rope with nine strands, called "cable laid."

Cable-Chain. Chain attached to the anchor.

Cable Stopper, (or Chain Stopper). A large casting for holding the anchor chain when anchor is in use.

Camber. The athwartship curvature of a deck. Sometimes called " round up."

Cant Frame. A frame not square to the keel line (small frame supporting overhang of stern).

Capstan. A revolving device, with axis vertical, used for heaving in lines.

Cargo. The freight carried by a ship.

Cargo Batten. Strips of planking on the inside of the frames in the hold to keep cargo away from shell of ship.

Cargo Boom. Heavy boom used in loading cargo.

Cargo Hatch. Large opening in a deck to permit loading of cargo.

Carling. Fore-and-aft member at side of hatch, extending across ends of beams where cut to form hatch.

Casing. Trunk enclosing portion of a vessel.

Caulk. To make water-tight or oil-tight.

Caulker. A man who makes seams tight in wood or metal. Caulker usually means a metal caulker. If wood caulker is meant the term " wood caulker " is used.

Caulking Tool. A blunt chisel (or other shape) for caulking joints.

Ceiling. Wood or steel sheathing on inner bottom frame brackets to protect cargo. Solid planking which is placed across the plating of the Tank Top to protect it from the cargo.

Cellular Double Bottom. Space between tank top, or inner bottom plating and outer shell of a vessel, divided by both longitudinal girders and transverse floors into separate compartments or cells.

Center Line. The middle line of the ship, from stem to stern.

Center Punch. A small, round hand punch used to mark the center of a hole.

Chain. Usually refers to heavy chain attached to anchor.

Chain Locker. Compartment in forward lower portion of ship in which anchor chain is stowed.

Chain Pipe. Pipe for passage of anchor chain from deck of ship to chain locker.

Chart House. Small room under bridge used for charts and navigation instruments.

Checkered Deck Plating. Cast or blind punched plates have diamond-shaped projections used for deck and ladder plates.

Chock. Cradle or support to prevent objects from shifting, usually of wood. A heavy fitting through which ropes or hawsers may be led.

Chock, Bower. An extra large chock, fitted well fordward for use of bower anchor.

Chock, Roller. One with a sheave or sheaves to prevent chafing of lines.

Chute, Slop. A trunk for carrying the slops down the side and discharging them into the water just above the waterline.

Cleat. A fitting attached to the deck, having two fore-and-aft arms or projections around which a rope or line may be secured.

Clip. A short angle bar.

Coaming. The vertical plates of a hatch or skylight, which extend above the deck.

Cofferdam. The space between two bulkheads located very close together.

Collar. A flanged band or ring.

Colliers. Vessels designed to carry cargoes of coal.

Collision Bulkhead. First water-tight bulkhead from bow of ship.

Compartment. A subdivision of space or room in a ship.

Compensation. The adding of "doubling plates" around hatchways, etc., to give local strength.

Composite Vessel. One, the framework of which is made up of steel angles and wood planked hull and decks.

Condenser. Apparatus in the engine room to liquefy the exhaust steam.

Corrosion. The oxidation of iron in water forming, first as a light red coat and progressing to a deep red rust scale.

Counter. Overhang at stern of ship.

Counterplates. Shell plates around the stern at the upper or weather deck.

Countersink. The taper of a rivet hole for a flush rivet.

Cowl. Hood-shaped top of ventilator pipe.

Cradle. A form often used in the plate shop on which a plate is shaped by the furnace man.

Crimp. Same as Joggle.

Cross Bunkers. Coal bunkers running athwartships.

Crow's Nest. A perch or platform at the mast head for lookout or signal man.

Cruiser Stern. A rounded, canoe shaped counter of the ship.

Cut Water. Portion of stem at water line.

D

Davit. Heavy vertical pillar, of which the upper end is bent to a curve, used to support the end of a boat when hoisting or lowering.

Davit, Anchor. A davit for "catting" the anchor.

Davit, Welin. A patent davit which swings the life boat out on a rocker arm instead of rotating about a vertical axis.

Dead Flat. The flat part of the amidship side or bottom of a ship.

Dead Rise. The rise or upward slant of the bottom of a ship from the keel to the bilge.

Deadweight. The total weight of cargo, fuel, water, stores, passengers and crew, and their effects, that a ship can carry.

Deck. The part of a ship that corresponds to the floor of a building.

Deck (Awning). Shelter deck.

Deck (Boat). One on which life boats are carried.

Deck (Bridge). Partial deck extending from side to side of ship, about amidships.

Deck (Forecastle). Partial deck at bow of ship, raised above upper deck.

Deck (Forecastle, Sunk). Is one which is partly above and partly below the Upper Deck.

Deck (Girder). Strength member extending either fore-and-aft or athwartships to support deck.

Deck (House). Shelter built on exposed weather deck.

Deck (Hurricane). Or boat deck.

Deck (Lower or first). First full deck above tank top or bottom.

Deck (Main or second). Second full deck above tank top or bottom. Principal deck of the main part of the hull.

Deck (Poop). Partial deck at stern of ship, raised above upper deck.

Deck (Promenade). Set aside for use of first-class passengers on passenger ship.

Deck (Upper or third). Third full deck above tank top or bottom.

Deck (Orlop). Partial or balcony deck between tank top and lower deck.

Deck Height. The vertical height from the top of a deck to the top of the next deck above.

Deck Load. Cargo stowed on weather deck.

Deck Plan. A drawing showing the layout of a deck, either for plating or arrangement.

Deck Stringer. The strake of plating that runs along the outer edge of a deck.

Decking. Wood, canvas, or any other material used to cover a deck.

Depth. The distance between the top of the Upper Deck beam, at the side, to the base line of the vessel, measured half way between the ship's forward and after perpendiculars.

Derrick. A device for hoisting heavy weights, cargo, etc.

Diamond Plates. Diamond-shaped plates connecting the web frames to the side stringers and acting as brackets to stiffen the framework.

Dismast. To remove the masts from a ship.

Die. A tool for forming a rivet head (applied to rivet dies).

(*a*) Flush die—to flatten rivets into a countersunk hole.
(*b*) Snap die—to form a round head.

Displacement. See Tonnage.

Distilling Apparatus. Machinery for distilling sea water to provide fresh water for the ship.

Docking Keel. A keel fitted on both sides of a large ship, bottom of each on a level with outside of flat keel, used to distribute the weight of a ship when in dry dock. Commonly used on battle ships.

Dog. A small, bent metal fitting used to secure doors, hatch covers, manhole covers, etc.

Dolly bar. A heavy bar to hold against a rivet.

Donkey Boiler. Small boiler used to drive winches, deck and auxiliary machinery.

Double Bottom. Compartments at bottom of ship between inner and outer bottoms, used for ballast tanks, water, fuel, oil, etc.

Doubling Plate. A plate fitted outside or inside of another to give extra strength or stiffness.

Draft. The vertical distance of the lowest part of the ship below the surface of the water when she is afloat.

Draft Marks. These are figures six inches high, placed six inches apart. They are located at bow and stern and denote the draft for every foot between the "light load" and "deep load" condition.

Draft, Mean. An average of the draft at the bow and stern.

Drag. The amount that one end of the keel is below the other when the ship is afloat, but not on an even keel.

Drain Holes. Small holes through floor plates to drain water toward the suction pipe.

Drift. To force two rivet holes into line with each other by use of a drift pin.

Drift Pin. A small tapered tool used to draw adjoining parts together, so that the corresponding rivet holes will come in line.

Dry Dock. A floating (or stationary) box-shaped structure into which vessels may be towed and the water pumped out from around the vessel for purposes of investigation or repairs of the ship.

Drying Room. A space just over Boiler Room and next to uptake. Used to dry linen, crew's washing, etc.

E

Engine Seating. The foundation for engines.

Entrance. The forward part of the hull at the water line.

Equipment. Anchors, cables, tow ropes, warping, mooring lines, etc., which are specified by the registration society under which the ship is being built, according to the ship's "equipment number."

Erection. The process of hoisting into place and bolting up the various parts of the ship's hull.

Even Keel. A ship is said to be on even keel when the keel is level, or parallel to the surface of the water.

Expansion Plan of Shell. A drawing having the correct position, size and mark of every shell plate.

Expansion Trunk. A built-up trunk extending fore-and-aft, over a series of oil compartments, of sufficient capacity to admit the expansion of oil due to changes of temperature.

Eye Bolt. A bolt formed with an eye or ring at one end.

F

Fabricate. To punch, cut, sheer, drill, bend, flange, or weld plates and shapes.

Fair. Smooth without abruptness or unevenness, in agreement. Fairing the lines consists in making them smooth. Rivet holes are fair when they agree one with another in adjoining members.

Fairlead. A small fitting through which a rope, or chain may be led so as to change its direction without excessive friction.

Fantail. Term applied to the whole of the round, overhanging stern aft of the transom frame.

Faying Side. The side of a plate which the punch enters.

Faying Surface. The surface between two adjoining plates.

Felt. Used in thin cakes as insulation material.

Fire Main. A large pipe used only to supply water to the fire hydrants from the fire pump.

Fender or Bumper. A fitting or device to prevent damage to a ship's hull at or near the waterline by other vessels, floating objects, docks, etc.

Fidley Hatch. Hatch around smokestacks and uptake.

Fine Lines. When ship is sharp pointed at the ends.

Flagstaff. Flag pole at stern of ship.

Flange. Portion of a plate or shape at, or nearly at, right angles to main portion.

Flare. Curvature of the forward frames outward.

Flat. A small, partial deck, built level, without curvature.

Floor. The lower portion of a transverse frame, usually a vertical plate extending from center line to bilge, and from inner to outer bottom.

Flush Rivet. A rivet driven with a flush die. (Has a point even with outside surface of plate.)

Fore-and-aft. In line with the length of the ship, longitudinally.

Fore and Afters. The longitudinal bars beside a hatchway; supporting the ends of the half beams.

Forecastle. The forward, upper portion of the hull, generally used for the crew's quarters.

Fore Peak. A large compartment or tank just aft of the bow in the lower part of the ship, used for trimming ship.

Forefoot. Bottom of the curve of the stem.

Forging. A mass of steel worked to a special shape by hammering while red hot.

Forward. Near or toward the bow.

Foundations. Supports for engines, boilers, etc.

Frame Mold. Template for the frame of a ship.

Frame, Side. Inside frames of a ship connecting to shell plating.

Frame Spacing. The fore-and-aft distance between adjacent frames.

Frames. Upright members or ribs forming the skeleton of a ship.

Frames, Continuous. Frames combining side frames and floors, extending continuous through decks.

Frames, Reverse. An angle fitted on the back side of the original frame to reinforce same.

Frames, Web. Heavy built-up frame of angle iron and plates.

Framing. The support and stiffening of the shell plating, deck plating. Usually consists of the ordinary transverse frames (or " ribs ") beams, floors, and the longitudinal framing, as keel, keelsons, longitudinals and stringers.

Freeboard. The vertical distance from the upper watertight deck or top of bulwarks to waterline, when ship is fully loaded.

Freeing Port. Holes through bulwarks of ship to discharge water from deck in case of shipping a sea.

Full Lines. When ship is blunt pointed at the ends.

Funnel. Smokestack of a ship.

Furring. Strips of wood bolted to frames of ship to hold cabin or storeroom lining. Also on tank top to hold wood planking.

G

Galley. Cook room or kitchen of a ship.

Galley Dresser. A cook's work table.

Galvanizing. Coating metal parts with zinc for protection from rust.

Gangway. A passageway, a ladder or other means of boarding a ship.

Garboard Strake. See "Strake."

Gauge. The distance from the center of a rivet to heel of an angle or edge of the plate.

Gauge Line. The line along which rivets are located in an angle or beam.

Girder. Strength member used as a support.

Girth. The transverse distance around a ship.

Gravity Tank. See Tank.

Grommet. A thread of oakum dipped in thick paint wrapped around a bolt before tightening, to make it water-tight.

Gross Tonnage. See Tonnage.

Ground Ways. Timbers fixed to the ground, under the hull on each side of the keel, on which she is launched.

Gudgeons. Bosses on stern post drilled for pins, on which rudder swings.

Gun. A name applied to a compressed air riveting hammer.

Gunwale. The side of a ship above the weather decks.

H

Harpin. A curved wooden piece used to hold frames at ends of ship in position when first erected.

Hatch. Opening in deck, for passage of people or cargo.

Hatch Batten. Narrow strips, usually of flat iron, which are fitted tight against the coamings to hold the canvas hatcb covers.

Hatch Beam. See Beam.

Hatch, Booby. A steel hood used to protect a hatchway for a ladder or stairway.

Hatch, Cargo. Large openings in a deck to permit loading of cargo.

Hatch Coamings. The vertical plates around the edge of a hatch-way.

Hatch Cover. Consists of wood planks, or steel plating.

Hatch, Fidley. Hatch over Boiler Room.

Hatchway. The vertical opening under a hatch.

Hawse Pipe. Casting extending through deck and side of ship for passage of anchor chain.

Hawser. A large rope.

Hawser Hole. Hole through bulwark for passage of a rope.

Heater or Heater Boy. One who heats rivets.

Heating Tongs. Tongs used to take a rivet from the fire.

Heeling. The degree of inclination of a vessel from the perpendicular.

Helm. The direction to which the tiller is put, or opposite to which the rudder is put. (When the rudder is to port the ship is said to carry starboard helm.)

Hog Frame. A long, heavy, curved girder used on the large, shallow draft river steamers. The ends of the vessel are thus supported by the middle part of the hull.

Hog Sheer. When a vessel is designed and built with the ends lower than the middle, it is said to have a "Hog Sheer." Many car floats used for transporting heavily loaded freight cars are built with a Hog Sheer.

Hogging. Straining of the ship that tends to make the bow and stern lower than the middle portion.

Hold. That part of a vessel where cargo is carried.

Hold Beams. Beams in a hold, similar to deck beams, but having no plating or planking on them.

Holder-On. See Riveting Squad.

Holding-down Bolts. The bolts which hold any machine to its seating.

Hook Stick. A rig used to hold drilling machine for light drilling.

Horn. To line or square up.

Hull. The body of a ship, including shell plating and frames.

Hydraulic Ram. A piston moving in a cylinder which operates by water power and is capable of exerting a tremendous pressure. Often used to start a ship in launching.

Hydraulic Riveter. See "Riveting."

I

In and Out Plating. A system of staggering strakes of plating by keeping one plate inside (inner strake) and the next plate outside (outer strake); then filling the space under the outer strake with a parallel liner.

In Way of. A nautical term meaning "near-by" or "parallel to."

Inboard. Inside the ship, toward the center line.

Inner Bottom. Plating forming the upper boundary of the double bottom. Also called " tank top."

Inserted Packing. Canvas strips soaked with red lead fitted between connections that cannot be caulked successfully, to insure a tight job.

Insulation. Asbestos covering often put over steam pipes to retain the heat.

Intercostal. Made in separate parts between frames, beams, etc., the opposite of continuous. (Floors are continuous; longitudinals intercostal, in a transverse framed merchant ship.)

Isherwood System. System of framing a ship chiefly with longitudinal members.

J

Jackstaff. Flag pole at bow of ship.

Jam. A special type of holding-on hammer used in heavy riveting.

Joggling. Offsetting the edges of plates of outer strakes to avoid the use of " liners."

K

Keel. The fore-and-aft member, usually in the form of horizontal flat plates end to end, extending on the center line from stem to stern along the bottom of a ship.

Keel, Bilge, or "Rolling Chock." Narrow plate and angles, fitted to reduce rolling motion of ship.

Keel Blocks. Heavy blocks on which ship rests during construction.

Keel, Vertical (Keelson). One made up of vertical plates extending fore-and-aft, located on the center line of flat keel.

Keelson, Side. Fore-and-aft member located on each side of center keelson.

Knee. A bracket connecting a beam to the frame. See "Beam Knee."

King Post. Vertical post to support cargo booms.

Knuckle. A sharp bend in a plate or bar.

Knuckle Plate. One with a sharp bend.

L

Ladder. Inclined steps, aboard ship taking the place of " stairs."

Landing Boards. Boards on deck beside hatches, for landing cargo when loading.

Landing Edge. The edge of a plate which is nearest the landing.

Lap or Landing. A joint, in which one part of a plate overlaps another, thus avoiding the use of a butt strap.

Launching. The operation of placing the hull in the water by having it slide down the launching ways. During launching the weight of the hull is borne by the " sliding ways," which are attached to the hull and slide with it down the " ground ways."

Pivoting. A condition which a ship is liable to assume if the water is too high over the ends of the ground ways. The after end of the ship will float before the bow has traveled far and the weight of the ship would be water-borne aft while the weight of the forward end of the vessel would come on the "fore poppets."

Tipping. This is liable to happen if there is an insufficient depth of water over the ends of the ground ways. As soon as the center of gravity of the hull had passed the ends of the ways, the stern of the ship would drop into the water, and the bow would rise from out of the cradle—a dangerous condition, as the cradle would collapse before the bow returned to it and then the cradle would be of no use in supporting the ship.

Launching Grease. Melted Russian tallow of the best quality smeared over with soft soap and train oil to increase lubrication is one of the many combinations on the market.

Laying off. Marking plates, shapes, etc., for shearing and punching from template.

Lay Up. To strike the plate near a rivet hole after the hot rivet has been inserted, in order to make the head of the rivet lay closely to the plate.

Length between Perpendiculars. The distance from the fore part of the stem to the after part of the stern post on a line of the upper deck.

Length over All. The length of a ship measured from the stem to the aftermost point of the stern.

Lift. To lift a template is to make it from measurement given and also to suit conditions.

Life Boats. Small boats stowed on the weather deck for use in case of an accident or sinking of the vessel.

Light, Deck. Small, thick glass in a frame which is let into weather decks for light to compartments below.

Lightening Holes. Holes cut in plates and frames to reduce weight.

Light Shaft (or trunk). Vertical trunk with glass covers to provide light to the compartment below.

Lights.

(*a*) **Mast Head Light.** A white light carried on the foremast from 20 to 40 feet above the hull.

(*b*) **Side Lights.** Port light or starboard light located on each side of the ship. (Red on port and green on starboard side.)

(*c*) **Towing Lights.** Two extra white lights, one above the other and not more than 6 feet apart to indicate that the ship has a tow.

Limber Chain. A small chain through the limber holes in the floors which is pulled back and forth to keep holes free of dirt and allow passage of water.

Limber Hole. A hole of a few inches diameter cut in a floor plate near the bottom to allow water to drain through it.

Liner, Parallel. A short, narrow plate fitted between an "outside" plate and beam, or frame (used as a distance piece).

Liner, Tapered. A small tapered plate fitted between plates at a lap.

Lines. The plans of a ship that show its form. From the lines, drawn full size on the mold loft floor, are made templates of the various parts of the hull.

List. As applied to the position of a ship in the water. Due to the unequal loading of the vessel, having more weight on one side of the center line than on the other side. The ship takes a position with masts, etc., on one side of a vertical center line.

Loftsman. A man who lays out and makes molds for a ship.

Log. A device similar to a cyclometer or speedometer which is towed behind a ship to record the number of miles traveled. Also the name of the record, or diary, kept by the captain of a vessel.

Log Line. A rope used to tow the log behind a ship.

Longitudinal. A fore-and-aft vertical member running parallel, or nearly parallel, to the center vertical keel through the double bottom. In merchant ships longitudinals are often intercostal.

Louvre. An opening in a door or bulkhead, with sloped shutter plates to prevent observation and also serving to ventilate the compartment inside.

Lug. A short length of angle bar. (See Clip.)

Lugpad. A projection on deck with hole for fastening a block for a lead.

M

Main Deck. See Deck.

Machine Riveting. See "Riveting."

Manhole. A round or elliptical shaped hole cut in a bulkhead or tank top. Large enough for a man to pass through.

Margin Bracket. One connecting side shell to margin plate of inner bottom.

Margin Plate. Outboard strake of inner bottom plating connected to shell.

Marker. Brass pipe dipped in paint for marking rivet holes.

Mast. A large round piece of timber or iron tube standing nearly vertical, at the center line of ship on the deck.

Mast, Fore. The first mast from the bow of a ship.

Mast, Main. The second mast from the bow of a ship.

Mast, Mizzen. The third mast from the bow of a ship.

Mast, Pole. A mast made in one piece throughout its length.

Mast Step. A socket or fitting to hold the foot of a mast.

Mean Draft. "See Draft."

Merchant Ship. A ship designed to carry cargo. A freight ship.

Mess Room. Dining room for officers and crew of a ship.

Midship. At the middle of the ship's length.

Midship Section. A plan showing a cross-section of the ship amidships. This plan shows sizes of frames, beams, brackets and thickness of plating.

Mold. A light pattern of a part of a ship. Usually made of thin wood or paper. Also called a template, for laying out plates or shapes.

Mold Loft. A shed or building with large, smooth floor on which the lines of a ship can be drawn to full scale and templates lifted.

Molded Dimensions. The lines as developed on the mold loft floor.

Mooring. The act of securing a vessel to a particular place.

Mooring Pipe. An elliptical-shaped hole in the bulwark plating; protected by a rounded casting through which hawsers are passed.

Mushroom Ventilator. See "Ventilator."

N

Natural Ventilation. See "Ventilation."

Net Tonnage. See Tonnage.

O

Oakum. Material used for caulking seams of wood deck.

Oil Tight. Riveted and caulked to prevent oil leakage. (Rivets must be more closely spaced for this purpose than for watertightness.)

Old Man. A rig for holding a drill.

On Board. On or in the ship.

On Deck. On the upper deck, in the open air.

Open Bevel. See "Bevel."

Open Rail. A railing composed only of pipe rails.

Outboard. Direction out from center line of ship toward either side.

Outreach. The distance a derrick can reach beyond its mast.

Overall Length. Length beyond the distance between perpendiculars. (Total length.)

Overboard. Outside, over the side of a ship. Into the water.

Overhang. Portion of the hull over, and unsupported by, the water.

Overhead Cable System. A method for handling material, when building a vessel, from a traveling trolley which runs on a large steel cable over the vessel and supported at each end by large wooden or steel columns or masts.

Oxter Plate. Bent shell plate which fits around the upper part of the stern post. Sometimes called, " tuck " plate.

Oxy-acetylene Flame. A very hot flame used for cutting steel made by mixing about 3 parts of oxygen to 2 parts of acetylene gas. Also used in different proportions for welding.

Oxy-Acetylene Welding. See "Weld."

P

Packing. Material put between plates or shapes to make them watertight; planking or wooden blocks supporting ship on sliding ways, previous to launch; steel washers fitted under the nut when bolting-up.

Pad Eye. An eye located on deck for fastening cables.

Panhead Rivet. A rivet with a pan-shaped head.

Panting. The in and out movement of shell plates at the bow of a ship, when at sea, due to water pressure when the ship is facing a heavy gale.

Panting Stringers. Fore-and-aft members on inside of shell at bow to prevent panting.

Parallel Body. Same as "Dead Flat."

Passer. A man who passes rivets to the holder-on, and often puts them into the rivet hole.

Passing Tongs. Tongs used in passing a rivet,

Pattern. A wooden piece made exactly like the finished casting, with allowance for shrinkage, which is put into the mold and forms a cavity into which the iron or steel is poured.

Pay. To pour hot pitch into a joint after it has been caulked. Also used to mean to "let out" or loosen ropes, etc.

Pillar. Vertical member or column giving support to a deck. Also called stanchion.

Pintle. Fitting, or pin on the rudder which turns in a gudgeon.

Pitch. The distance apart of rivets on the same line; the up and down motion of a ship; material used to make water tight joints between wood planks.

Plan. A drawing prepared for use in building a ship.

Planking. Wood covering for decks, etc.

Plate Bending Rolls. Three large adjustable steel cylinders, two below and one above, used to bend or flatten plates, etc.

Platform. A partial deck.

Plating. The plates of the shell, a deck, a bulkhead, etc.

Plimsoll Mark. This is a figure, with letters, which is located and painted on the outside of the side shell, directly amidships and at a height varying with the type of ship. Done to determine the safe freeboard of the vessel.

Poop. The after, upper portion of the hull, usually containing the steering gear.

Poppet. The cradle holding the bottom of the ship on the sliding way at forward and after ends during launching.

Port. The left-hand side of the ship when facing the bow.

Port, Blind. A port with a steel door which looks like the shell plating when the port is closed.

Port, Cargo. A door cut into the side of a vessel for loading and unloading cargo.

Port, Coaling. These are fitted in the side shell plating on some types of vessels whose voyages necessitate their coaling in foreign ports where it is customary for ships to be coaled through the side rather than down through the deck.

Port, Freeing. An opening in the bulwark plating to allow water to run overboard.

Porthole ("Sidelight"). A circular opening in the ship's side for light and air.

Propeller. Wheel consisting of two or more blades for propelling a ship.

Propeller Arch. The arched part of the hull just over the propeller.

Propeller Shaft. See "Shaft."

Pump. See chapter on "Piping Systems."

Punch. A machine for punching holes in plates and angles.

Q

Quadrant. A fitting on the rudder head to which the steering chains are attached.

Quarter. A side of the stern.

Quarter Pillars (or Stanchion). A line of pillars about half way between the center line and the outside of the ship. Needed on ships of wide beam.

Quarters. Living or sleeping rooms for officers and crew of a ship.

Quick Closing Gear (or Long Arm System). A device for closing bulkhead doors quickly from the wheel house in case of bilging.

R

Rabbet. A depression or offset designed to take some other adjoining part; as, for example, the rabbet in the stem to take the shell plating.

Rag. The excess metal sticking above the plate on the end of a rivet after driving and before it has been "chipped."

Rail. The upper edge of the bulwarks.

Rake. Slope of the bow when tilted forward. If the stem was built plumb, it would appear to tilt aft, because of the downward curve of the sheer.

Rally. A period of 4 or 5 minutes of continuous work before a rest. A term used when launching.

Reaming. Enlarging a rivet hole by means of a revolving cylindrical, slightly-tapered tool with cutting edges running along its sides.

Refrigerating Machine. A machine for keeping refrigerators cold, operating by either the cold air, dry air, or ammonia system.

Reverse Frame. An angle bar or other shape riveted to the inner edge of a transverse frame to reinforce it.

Ribband. A fore-and-aft wood strip or heavy batten used to support the transverse frames temporarily after erection.

Rider Plate. The center line strake of plating on the Tank Top.

Rigger. Men hoisting material and doing rigging around the yard. Rigging ships with ropes and awnings, etc.

Rigging. Manila and wire ropes, lashings, etc., used to support masts, spars, booms. Also the handling and placing on board the ship of heavy weights and machinery.

Rigol (Eyebrow). This is a small watershed in the form of an angle which is riveted to the shell just above an air port and is bent to curve around it; to prevent water from dripping down across the glass.

Ring Bolt. An eye bolt having a loose ring in the eye, for lifting hatches, manholes, etc.

Rise of Floor. The amount that the flat portion of the bottom of the ship rises from the keel to the side of the ship.

Rivet. A short steel bolt usually driven or clinched after being heated red hot and used to fasten plates, or plates and angles together.

Rivet Heads. See "Notes on Riveting," Chapter II.

Rivet, Loose. A term applied to a rivet which was not driven tight and rattles when it is hit by the hammer.

Rivet Passer. The boy who catches the rivets in a tin cone after they have been thrown by the rivet heater, and puts them into the rivet hole, or who passes the rivet to the Holder-on.

Rivet Squad. Usually consisting of four men, the rivet heater, the rivet passer, the bucker-up and the riveter.

Bucker-Up. The man who holds a dolly bar or air jam up against the head of a rivet while the point is being riveted over.

Holder-On. Same as Bucker-Up.

Rivet Tack. These are used near the center of a plate, when two or more plates are superimposed and fastened together, to hold the plates tightly together on all parts of the surface.

Riveted Joints:

Single Riveted. One line of rivets.

Double Riveted. Two lines of rivets.

Treble Riveted. Three lines of rivets.

Quadruple Riveted. Four lines of rivets.

Chain Riveting. Two lines of rivets directly opposite each other.

Staggered Riveting. Two lines of rivets with spaces alternated.

Butt Joint. A joint formed by butting the edges of the plates together and covering with one or two straps.

Lap Joint. A joint formed by lapping one plate over the other without using any strap.

Riveter. The man who forms the point on the rivet with a pneumatic hammer or hand hammer.

Riveter, Hydraulic. A large machine with U-shaped frame which drives rivets by water pressure.

Riveter, Pneumatic. A compressed air hand tool for riveting.

Riveting, Hand. To form the point of a rivet by using a hand hammer.

Riveting, Machine. A term applied to any system of riveting other than hand. Generally refers to pneumatic riveting.

Roll. Motion of the ship from side to side alternately raising and lowering each side of the deck.

Rose Box. Screen around the end of a bilge suction pump.

Rough or "Service" Bolt. Used to bolt a plate or frame to ship.

Round Up. Same as "Camber."

Rudder. A large fitting hinged to the rudder post. Used for steering the ship.

Rudder Arms. Heavy steel arms forming a part of the rudder stock and riveted to the rudder plate.

Rudder, Balanced. A rudder having a portion forward of the stock and a portion aft of it.

Rudder Pad Eye. A pad eye attached to the top of the rudder so that in case all the steering apparatus breaks down, cables called relieving tackles may be attached to the rudder through this eye and led up over the sides of the ship to be operated by hand.

Rudder Plate. The vertical plate of a rudder.

Rudder Post. The vertical post at after end of stern frame which supports rudder.

Rudder Stock. Shaft of rudder which extends up through ship and is connected to steering gear.

Rudder Stop. A fitting on rudder post to limit the swing of the rudder.

Rudder Trunk. A small water-tight vertical compartment directly over the rudder to allow the rudder stock to pass up through.

Run. The part of the hull, aft of the bilge, below and at the water line.

Rungs. The steps of a ladder.

S

Saddle. An A-shaped trunk or shaft through which coal is passed when filling the coal bunkers.

Sagging. Straining of the ship that tends to make the middle portion lower than the bow and stern.

Samson Post. A heavy vertical post that supports cargo booms.

Scantlings. The dimensions of various parts of the ship.

Scarph. To thin out a corner of a plate to allow for another plate to lap over it without extra thickness.

Scrap. Waste pieces of material.

Screen Bulkhead. One that is dust tight only.

Scrieve Board. A large section of flooring in the Mold Loft in which the lines of the body plan are cut in with a knife. Used for making molds of the frames, beams, floor plates, etc.

Scrieve Knife. A knife used to cut the lines (about $\frac{1}{16}$ in. wide and $\frac{1}{16}$ in. deep) into the scrieve board in the Mold Loft.

Scupper. A drain from the edge of a deck discharging overboard.

Scupper Pipe. One which drains water from scuppers to side of ship.

Scuttle. A small hatch.

Seam. Fore-and-aft joint of shell plating.

Seam Strap. Butt strap of a seam.

Serving Board. A tool used in the same way as serving mallet.

Serving Mallet. A tool used in serving or wrapping cord about a rope or cable.

Set Iron. Bar of soft iron used on bending slab to give shape of frames.

Shaft Bracket. A bracket supporting the shaft in a twin-screw vessel.

Shaft, Crank. This shaft is in the engine bed and connects to the cranks of a reciprocating engine.

Shaft, Tail. The short, round shaft connected to the propeller and extending forward through the stern tube.

Shaft, Thrust. This piece is fitted with thrust collars which engage with the thrust block.

Shaft, Tunnel. This is made up of five or six lengths and extends from tail shafts to the section at the "thrust block."

Shaft Tunnel or **Shaft Alley.** Enclosed passage around propeller shaft extending from engine room to after peak bulkhead.

Shape. Long bar of constant cross-section, such as a channel, T-bar, angle-bar, etc.

Shears. Large machine for cutting plates and shapes.

Sheer. Fore-and-aft curvature of a deck.

Sheer Plan. Side elevation of a ship's outline.

Sheer Strake. See Strake.

Shell Expansion. A plan showing details of all plates of the shell.

Shell Landings. Places on the frames showing where the seams of shell plates come.

Shell Plating. The plates forming the outer skin of the hull.

Shift. To change position on a ship.

Shifting Board. A fore-and-aft center line partition of portable boards to prevent bulk grain from shifting sideways.

Shift of Butts. To stagger the joints in shell or deck plating so that they do not occur in same straight line.

Shore. A large wooden support or prop.

Shore, Bilge. The same as Bottom Shores, only these are placed along the bilge of the ship.

Shore, Bottom. These are heavy wooden props or logs of wood, placed under the bottom plating to help steady and support the weight of the ship while building.

Shore, Dog. These are used to hold the sliding ways from moving toward the water until all is ready for the launch. Just before the "saw-off" plank is cut, these dogs are dropped out of place. They are laid on an incline and act as a brace.

Shore, Side. Similar to Bilge Shores and are placed against the sides of the ship, especially at the ends where there is considerable weight in the overhanging structure.

Shore Spur. A slanting brace on either side of ship or ways.

Shoulders, Bows. Where the bilge curve dies away. Both sides just aft of the stem. (Port bow—Starboard bow.)

Shoveling Flat. A wood platform at the bottom of a coal bunker opposite the coal door.

Shutter. A hinged steel plate.

Side. Portion of hull above the bilge.

Sidelight. A metal frame riveted in a hole in the shell of a vessel and carrying thick glass, to illuminate the inside of the ship.

Sidelight Screen. An open, boxlike affair, with bottom, one side and an end. One on each side of the ship, placed on the side of the bridge or deck house to act as a receptacle for the Running Light and so constructed as to prevent the light from shining across the bow of the ship.

Skylight. An opening in a deck to give light and air to the compartment below it.

Sliding Ways. (See Launching.)

Slop Chute. See Chute.

Smoke Stack. Large vertical funnel for passage of smoke and gases from boilers.

Snap Rivet. A rivet driven with a snap die. (Has a round or button head and point.)

Soft Packing. An oily or tarry material placed under plates or angles to make them watertight.

Soil Pipe. A drainage pipe leading from toilets, sinks, etc., to a storm valve in the ship's side just above the water line.

(NOTE.—A "storm valve" is one with a non-return flap so that the sea water cannot rush up the pipe when the ship is in a heavy sea. Passage through the pipe can be in only one direction.)

Sounding Pipe. Vertical pipe in oil or water tank used to measure depth of liquid in tank.

Spar. A long, round, wooden timber, or steel tube.

Spectacle Frame. A heavy steel casting having two arms on each side of the counter of ship. The boss for supporting the after end of the propeller shaft on a twin-screw ship is held in position by means of the two arms which are thoroughly secured to special framing inside the hull.

Stability. Tendency of a ship to remain upright.

Stage Bent. Upright support for staging.

Stage Planks. Timbers making up the floor of a staging.

Stage Uprights. The vertical members of a staging at the side of a vessel on the ways.

Staging. Timber platforms built up inside and around the outside of a ship during construction.

Stanchion. An upright member used as a support.

Stapling. Collars, forged of angle bars, to fit around continuous members passing through bulkheads or decks for watertightness.

Starboard. The right-hand side of the ship when facing the bow (opposite to " port ").

State Rooms. Private sleeping rooms.

Stay. A rope which stiffens and supports a mast or funnel, usually of steel.

Stealer. A strake of shell plating that is made extra wide to fit the width of two narrow strakes near the ends of the ship, where the girth is less.

Steerage. Is that portion of the ship having the poorest quarters for passengers. Generally the after part of the ship, due to the noise of the screw and the excessive motion when the ship is in a seaway.

Steering Gear. Apparatus for controlling the rudder.

Stem. Forging or casting forming extreme bow of ship, extending from keel to weather deck.

Step. To set in place as applied to a mast.

Stern. After end of ship.

Stern Frame. Large casting attached to after end of keel to form ship's stern. Often includes rudder post, stern post, and aperture for propeller.

Stern Tube. Tube in stern post through which propeller shaft passes.

Stiffener. An angle bar, T-bar, channel, etc., used to stiffen and form a support for plating of a bulkhead.

Stock. The name applied to material before it is laid out and cut.

Stokehold. Space in front of boilers in Boiler Room where the men stand when stoking the boilers.

Stool. Support for shaft bearings.

Stopwater. A piece of packing of canvas and red lead or other material to prevent leakage where caulking is impractical.

Storm Valve. A check valve in a pipe opening above water line on a ship.

Stow. To put away securely.

Stowage. Space for storing cargo, supplies, etc.

Strake. A fore-and-aft course or row of shell or deck plating.

Strake, Bilge. Shell plating at the curve of the bilge of the ship. (This term applies only amidship, where the bilge occurs.)

Strake, Garboard. The strake of shell plating on each side of the flat keel.

Strake, Margin. The outer strake of plating on the tank top.

Strake, Sheer. The strake of shell plating at the topmost, continuous deck (showing the "sheer" or curvature of the profile of the ship).

Strap. An extra plate put over a butt joint to transmit the strength from one plate to the other.

Stringer Plate. The outer strake of deck plating, as " main deck stringer." A fore-and-aft continuous member used to give longitudinal strength. Others are made up of a narrow plate and angles, riveted to shell, cut out around frames, runs " continuous," and according to location are called " hold stringers," " bilge stringers," " side stringers."

Strong Back. Bar for locking cargo port doors and water-tight scuttles.

Strut. Support for propeller tail shaft used on boats with more than one propeller.

Stud. A piece of iron fitted in the short axis of a link of a chain to prevent the chain from "kinking."

Stud Link. An elliptical-shaped link in a chain, having a short stud fitted between its sides.

Superstructure. Deck houses, etc., above main deck.

Swash Plate. Plates in oil or water tank to prevent excessive movement of the liquid (form a baffle plate).

Symbols (marks of identification). Following partial list is illustrative:

V.K.FL.C.	Vertical keel floor clip.
FL.FR.	Floor frame.
FL.S.	Floor stiffener.
BB.FL.C.	Bilge bracket floor clip.
S.D.B.B.	Second deck beam bracket.
E.C.UDK.B.C.	Engine casing upper deck beam clip.
CK.	Countersink.
CK.T.S.	Countersink this side.
CK.O.S.	Countersink other side.

T

Tail Shaft. See Shaft.

Tank, Daily Supply (Gravity Tank). Is located on the weather deck or on top of a deck house. Water is pumped there from the fresh-water tanks in Engine Room and it is then allowed to run by gravity, out of the tank as needed in the different state rooms and galley.

Tank, Fresh-water. This is the main supply for the water for drinking, cooking, bathing and is often held in one or two tanks located in one side of the Engine Room.

Tank, Reserve Feed Water. The space in the Inner Bottom, generally under the Engine Room, for carrying the extra water needed in the boilers for generating steam.

Tank Top. Plating over the inner bottom.

Tank, Topside. These are longitudinal compartments running fore-and-aft, just under the Upper Deck and extending out to near the side of the cargo hatches. They are built-in tanks used in ships which are to carry a "bulk" cargo as grain or coal. For such cargoes this space in the upper corners is often not used due to the fact that the cargo will not run there of its own accord.

Tanks, Auxiliary. For carrying lubricating oil; and other such small tanks.

Tap Bolt. A bolt threaded all the way up to its head.

Telegraph. Means of signaling from Bridge to Engine Room.

Telemoter. A device for operating the steering gear by water pressure run through small pipes from the steering engine to the wheel on the bridge.

Tell-tale Pipe. A small pipe leading from the ballast tanks into the Engine Room so the engineer can know when the tanks are full.

Template (Temporary Plate). Full-sized pattern.

Thrust Block. Solid casting with rings which fit between rings on one section of the tunnel shaft to take the thrust or pressure in a forward direction due to action of the propeller.

Thwart. Horizontal cross timbers in a staging.

Tie Plate. Narrow plates attached to deck beams under wood deck to give extra strength.

Tiller. Arm attached to rudder head for operating rudder.

To Trim Ship. To shift ballast to make a ship change its position in the water.

Toggles. Wood timbers around which rope is knotted or fastened.

Tomahawk. A tool used in backing out rivets.

Tonnage, Cargo. The basis for a cargo ton is generally taken as being forty cubic feet per ton; with 2240 pounds per ton weight.

Tonnage, Deadweight. Is the amount the ship can carry of cargo, stores and bunker fuel, at 2240 pounds per ton. It is the difference when the ship is in a "light" condition and when she is in a "loaded" condition.

Tonnage, Displacement. This is the weight of a complete ship with all that is in it at the time the estimate is made. This varies according to whether there is cargo aboard or the ship is "light." The word displacement is obtained from the fact that the ship will displace a volume of water whose weight is equal to the weight of the ship and contents.

(The average size freight steamer would have about the following)

Gross Tonnage	6,000
Net Tonnage	4,000
Deadweight Tonnage	10,000
Displacement, loaded	13,350

Tonnage, Gross. This term applies only to the size of a vessel, not to the cargo. This amount is determined for each vessel by dividing by one hundred the number of cubic feet of closed-in space in the hull and deck houses. A vessel ton is equal to one hundred cubic feet.

Tonnage, Net. This expresses the space available for the accommodation for cargo and passengers. It is obtained from the Gross Tonnage after deducting space for crew, engine room, fuel, navigation instruments, etc.

Transom. Rounded part of stern aft of the stern post.

Transom Beam. A beam across a transom frame.

Transom Frame. Aftermost main frame of ship attached to the stern framework; to which is attached the "cant frames."

Transverse. Athwartships, at right angles to the keel.

Transverse Frames. Athwartship members forming the ship's "ribs." (Above the floors.)

Trunk. Small casing passing between decks, such as is sometimes used for ladders.

Tumble Home. An inboard sloping of the ship's side above the level of greatest beam.

Tunnel Shaft. See Shaft.

Turnbuckle. A fitting with right- and left-hand threads on two rods which turn in a sleeve. Rotation of the sleeve draws the two rods together. A hook or eye in the outer end of each rod provides a place for attaching it to the ship structure.

Turret-Deck Ships. Are those with the upper sides carried in toward the center line and a narrow deck running fore-and-aft to connect the Forecastle Deck and the Poop Deck and deck house.

Twin Screw. Having two propellers.

U

Unfair Holes. Rivet holes which do not line up when the plates are fitted, one on the other.

Unship. To detach from its place in the ship.

Upset. To hammer on the end of a steel or iron rod while it is red hot, thus making it a larger diameter.

Uptake. Connection between boilers and smoke stack.

V

Ventilating Shaft (or trunk). An opening for fresh air, from the weather deck to a lower compartment in the ship.

Ventilation. The supply of fresh air to all living quarters in the ship.

Ventilation, Mechanical. Circulating fresh air in a ship by means of fans or blowers.

Ventilation, Natural. Supplying fresh air to certain parts of the ship by the natural tendency of hot air to rise or by the wind blowing past the ship.

Ventilation Outlet. An opening to exhaust foul air.

Ventilator. A device for furnishing fresh air to compartments below decks.

Ventilator, Cowl. A large funnel-shaped trunk capable of being swung in any direction toward the wind for a supply, away from the wind for an exhaust.

Ventilator, Goose-neck. A ventilating pipe with its upper end turned down toward the deck.

Ventilator, Mushroom. A ventilator pipe in a deck having a mushroom-shaped hood over it, which can be screwed down during stormy weather, and which serves to act as an air supply or exhaust.

Vertical Keel. Row of plating extending vertically along center of flat plate keel. Sometimes called center keelson.

Vibration. The shaking of a ship due to the unbalanced motion of the engines.

Voice Tube. Large speaking tube.

W

Warping Bridge. Bridge at after deck house used while docking a ship.

Water Ballast. Water let into some of the compartments of the Inner Bottom to act as a dead weight or ballast, when the ship is in a "light trim."

Water Course. This applies to any channel for passage of the water. Is generally spoken of along the line of limber holes in the floors.

Water Line. The line of the water's surface when the ship is afloat.

Waterline, Light. The line along the hull of the surface of the water when the ship is in a "light" condition, without cargo.

Waterline, Load. Line of the surface of the water when the ship is deep in the water due to the weight of the cargo aboard.

Water Logged. The condition of wood or other material which has become so soaked as to be heavier than an equal volume of water and hence will sink beneath the surface.

Watertight. So riveted or caulked as to prevent the passage of water.

Water-tight Flat. Section of horizontal plating, across the bow or stern of a ship, usually made into a tank top.

Waterway. A narrow passage along the edge of the deck for the drainage of the deck.

Ways. Timbers, etc., on which a ship is built or launched. (See Launching.)

Weather Deck. A deck with no overhead protection.

Web. The vertical portion of a beam, the athwartship portion of a frame.

Web Frame. A frame with a deep web.

Wedges. Long, wood pieces used by the ship carpenters to "wedge-up" the ship just before it is to be launched.

Weeping. A term applied to a leaky seam or rivet point when the Inner Bottom compartments are being tested with water.

Welding. Making a joint of two metal parts by fusing more metal in between them.

Welding, Butt. Is done when two pieces are put end to end and then welded together.

Welding, Electric. Heat is generated by means of an electric current. A positive "pole" is attached to one piece of metal and a negative "pole" to the other; the heat is sufficient to melt the metal so that when pressed together a weld is accomplished. Called "spot welding."

The other process is where an electrode of iron is used and an electric arc produced. The heat melts the electrode which is used to furnish the extra material for the weld. Called "arc welding."

Welding, Lap or Scarph. For this style of welding it is necessary to have one end of each of the two pieces drawn out on a long taper; these are fitted to match and then welded together.

Welding, Oxy-Acetylene. This is a process whereby the combination of oxygen and acetylene gases are used to produce an intense heat.

Welding, Thermit. In this work powdered aluminum, iron filings etc., are ignited and as they melt the weld is formed.

Well ("Sump"). Space in bottom of ship into which waste water runs, that it may be pumped out.

Well Deck Steamer. Is one with a Poop, Bridge and Forecastle Deck having a "well" both forward and aft of the Bridge Deck, formed by the bulkheads and the Upper Deck and side bulkwarks.

Whaleback Steamer. Is one that has the top sides carried in on a curve, the deck house being the only erection above the top of the hull. The engines and living quarters are generally located in the after end of the ship.

Wheel House (Pilot House). Is located on the Navigating Bridge and contains the steering wheel, compass and other navigating instruments.

Wild Cat. A wheel for the chain to ride around when being hove in by the steam windlass.

Winch. A small hoisting engine.

Windlass. The machine used to hoist the anchors.

Y

Yard. A horizontal, athwartship spar fitted to a mast.

Note.—We often speak of forward side and after side. It is, strictly speaking, incorrect, as it is the forward or after end; the other direction, at right angles, is port or starboard side, i.e., that part of a hatch nearest the bow is the forward end regardless of the size of the hatch.

Zeitfracht Medien GmbH
Ferdinand-Jühlke-Straße 7
99095 Erfurt, Deutschland
produktsicherheit@kolibri360.de